T0210918

Communications
in Computer and Information Science   679

Commenced Publication in 2007
Founding and Former Series Editors:
Alfredo Cuzzocrea, Dominik Ślęzak, and Xiaokang Yang

More information about this series at http://www.springer.com/series/7899

S. Subramanian · R. Nadarajan
Shrisha Rao · Shina Sheen (Eds.)

Digital Connectivity – Social Impact

51st Annual Convention
of the Computer Society of India, CSI 2016
Coimbatore, India, December 8–9, 2016
Proceedings

Editors
S. Subramanian
Karpagam Academy of Higher Education
Coimbatore
India

R. Nadarajan
PSG College of Technology
Coimbatore
India

Shrisha Rao
International Institute of Information
 Technology
Bengaluru, Karnataka
India

Shina Sheen
Applied Mathematics and Computational
 Sciences
PSG College of Technology
Coimbatore, Tamil Nadu
India

ISSN 1865-0929   ISSN 1865-0937 (electronic)
Communications in Computer and Information Science
ISBN 978-981-10-3273-8   ISBN 978-981-10-3274-5 (eBook)
DOI 10.1007/978-981-10-3274-5

Library of Congress Control Number: 2016958516

Printed on acid-free paper

This Springer imprint is published by Springer Nature
The registered company is Springer Nature Singapore Pte Ltd.
The registered company address is: 152 Beach Road, #22-06/08 Gateway East, Singapore 189721, Singapore

Preface

The most recent advancements in the dynamically expanding realm of Internet and networking technologies have provided a scope for research and development in computer science and its allied thrust areas. In this series, CSI 2016 organized by the CSI Coimbatore Chapter during December 8–10, 2016, invited submission of high-quality, original scientific papers presenting novel research, focusing on information and communication technologies (ICT) and generally all interdisciplinary streams of engineering sciences, having a central focus on "Digital Connectivity–Social Impact." It is an opportunity for researchers to meet and discuss solutions, scientific results, and methods in solving intriguing problems.

The theme "Digital Connectivity–Social Impact" was selected to highlight the importance of technology in solving social problems and thereby creating a long-term impact on society. The convention invites papers in four distinguished areas including computational intelligence, IT for society, network computing, and information science. Papers were solicited from industry, government, and academia (including students) covering relevant research, technologies, methodologies, tools, and case studies. The aim of the convention was to explore and emphasize the role of technology in real-world problems.

Computational intelligence (CI), a dynamic domain of modern information science has been applied in many fields of engineering, data analytics, forecasting, biomedicine, and others. CI systems use nature-inspired computational approaches and techniques to solve complex real-world problems. The widespread applications range from image and sound processing, signal processing, multidimensional data visualization, steering of objects, to expert systems and many other potential practical implementations. CI systems have the capability to reconstruct behaviors observed in learning sequences, and can form rules of inference and generalize knowledge in situations when they are expected to make predictions or to classify the object to one of the previously observed categories. The CI track consists of the research articles that exhibit various potential practical applications.

Information science is an interdisciplinary field primarily concerned with the analysis, collection, classification, manipulation, storage, retrieval, movement, dissemination, and protection of information. Information access is an area of research at the intersection of informatics, information science, information security, language technology, computer science, and library science. The objectives of information access research are to automate the processing of large and unwieldy amounts of information and to simplify users' access to it. Applicable technologies include information retrieval, text mining, machine translation, and text categorization; papers related to these topics were included in this track.

Network computing is a generic term in computing that refers to computers or nodes working together over a network. The broad term "network computing" represents a way of designing systems to take advantage of the latest technology and maximize its positive impact on business solutions and their ability to serve their customers. Network computing helps link organizations with their suppliers and customers across the world, brings the benefits of computing to new audiences, and extends the scope of electronic commerce. It encompasses cloud computing, distributed computing, and virtual network computing. This track highlights the networking capabilities needed to solve the most challenging problems in every domain.

Technology can be a powerful tool that can be harnessed to efficiently and effectively provide resources to those who need them. As technology spreads globally, the opportunity to use technology as a mechanism to solve pressing social problems grows. The goal of the "IT for Society" category was to stimulate new thinking on a broad range of social benefits of information technology.

We received 74 papers in total, and accepted 23 papers (31%). Every submitted paper went through a rigorous review process. Where issues remained, additional reviews were commissioned.

The organizers of CSI 2016 whole heartedly appreciate the peer reviewers for their support and valuable comments for ensuring the quality of the proceedings. We also extend our warmest gratitude to Springer for their continued support in bringing out the proceedings volume in time and for excellent production quality. We would like to thank all keynote speakers, Advisory Committee members, and the chairs for their excellent contribution. We hope that all the participants of the conference benefited academically and wish them success in their research career.

This CSI series traditionally results in new contacts between the participants and interdisciplinary communications, often realized in new joint research. We believe that this tradition will continue in the future as well.

December 2016

S. Subramanian
R. Nadarajan
Shrisha Rao
Shina Sheen

Organization

President CSI

Anirban Basu

Chair

P.R. Rangaswami

Program Chair

S. Subramanian

Program Co-chair

R. Nadarajan

Advisory Committee

Anirban Basu	President CSI
Nandini Rangaswamy	Chandra Group & Secretary GRG Institutions
C.R. Muthukrisnan	IIT-M, Consultant Advisor TCS
R. Chandrasekaran	Cognizant Technology Solutions
K. Ramasamy	Tamil Nadu Agricultural University, India
S. Subramanian	Karpagam Academy of Higher Education, India
S. Sundar Manoharan	Karunya University, India
P. Venkat Rangan	Amrita University, India
K. Ramasamy	Roots Industires Ltd.
B. Soundarajan	Suguna Foods Ltd.
G. Soundarajan	CRI Pumps Ltd.
Vijay Venkataswamy	Vantex Ltd.

Contents

Information Science

Texture Classification Using Shearlet Transform Energy Features

K. Gopala Krishnan[1]([⊠]), P.T. Vanathi[2], and R. Abinaya[1]

[1] Department of ECE, Mepco Schlenk Engineering College, Sivakasi, India
gopi1969@yahoo.com, abinayadevasena@gmail.com
[2] PSG College of Technology, Peelamedu, Coimbatore, India
ptvani@yahoo.com

Abstract. This paper presents a novel approach for texture classification using Shearlet Transform. The Shearlet Transform is a recently developed tool, which have the multiscale framework which allows to efficiently encode anisotropic features in multivariate problem classes. Shearlets are a newly developed extension of wavelets that are better suited to image characterization. In addition the degree of computational complexity of many proposed texture measures are very high. In this paper, a novel texture classification method that models the adjacent shearlet subband dependences. In this paper the classification efficiencies of Minimum Distance classifier was compared with SVM classifier efficiency. For texture classification, the energy features are used to represent each shearlet subband. Comprehensive validation experiments performed on different datasets proves that this research work outperforms the current methods due to efficient multiscale directional representation of Shearlet Transform.

Keywords: Shearlet transform · Subband dependence · Texture classification

1 Introduction

Texture is defined as the measure of variation of intensity of the surface determining properties such as roughness, smoothness, repeated, etc. Texture is generally classified into two types namely Surface texture and Visual texture. Surface texture is the primary visual cue observed in natural images. Visual textures are synthetically generated and has an isolated perceptual quality, simplified for study purpose only. Textures are the properties that appear on the surface of the objects, such as a person's fingerprint, repeated patterns on clothes, etc. The properties of the textures are very important that are visualized by humans and used in their daily life. These important features has caused that textures are inevitable and are involved in physical life applications, such as biomedical image processing, remote sensing, document processing, etc. Despite the lack of a universally agreed definition for texture, all researchers agree on two points. Texture analysis has several significant challenges due to the complexity of textural patterns and different lighting conditions that must be considered. It provide appropriate feature for subsequent studies, such as Image retrieval, pattern recognition, and image segmentation. Texture analysis consists of four major techniques namely Classification, Segmentation, Synthesis and Shape from texture [1]. In texture classification

S. Subramanian et al. (Eds.): CSI 2016, CCIS 679, pp. 3–13, 2016.
DOI: 10.1007/978-981-10-3274-5_1

process, the map for classification is drawn between the textures where each textured region is acknowledged with the texture class to which category it belongs. It is also said that, the goal of texture classification is to allocate an unidentified texture image to one of the set of identified texture class. The two main classification methods are supervised and unsupervised classification methods [2]. The supervised classification method is performed by using the trained set of textures to learn the behavior for each texture class. The unsupervised classification automatically discovers different classes from input textures that does not require any prior knowledge about the textures. Another classification method is semi-supervised, in which only limited preceding knowledge is present regarding the textures. These classification methods represented above consist of two stage process. Feature extraction is the first stage, where the description of each texture class in terms of features is measured. Identification and selection of different features that produces exact output are very important because they are invariant to inappropriate transformation of the image, such as translation, scaling, and rotation. The computable measures of certain values of features must be similar to the texture already stored in the database but it is difficult to design an universally acceptable feature extractor since most of the existing techniques depends on some predefined problems and it needs the basic knowledge about the working domain. The next stage is the classification phase, in which the features extracted from the training set are compared with the testing texture and the best fit is found. Texture classification process in general consists of two major phases:

A. *Learning Phase*:

Learning phase is the initial step in texture classification. The major objective of this phase is to create a model for texture content of every texture class stored in the database which is considered as the training set. The training dataset in general consist of texture images of known classes along with their labels. These texture contents stored for the training images undergo certain texture analysis methods. These analysis methods are used to extract the detailed information regarding the texture images which are denoted as features. These features obtained can be in the form of numbers or discrete histograms that are used to characterize the properties of the texture images. Some of the textural properties of the textures are contrast, smoothness, roughness, orientation, brightness, etc.

B. *Recognition Phase*:

Recognition phase is the major step in the process of texture classification where the textural features of the unidentified sample are defined with the same texture analysis method as that of learning phase. After obtaining the textural features of the training and testing samples, the features of training samples are compared with the features of testing samples by using the predefined classification algorithm. The comparison results provides the best match and if the match found is not good, the texture is said to be misclassified.

Texture analysis is the technique to obtain the detailed information of the textures and it is done by the following processes such as synthesis, classification, segmentation and shape from texture. Some of the applications of texture analysis techniques are identifying surface defects, medical diagnosis, ground classification and rain forecasting, text analysis, face recognition and fabric classification [3]. Textures are used to explain about the statistical and structural relationship between the pixels, and also it describes on the properties of the textures. Statistical method is mostly used for the natural textures that contains irregular surface patterns, for example grass, sand, bark surfaces. It is the measure of intensity arranged in the specified region. Likewise the structured method is used for the natural or artificially generated textures that contains repeated or regular surface patterns. Since feature extraction plays the major role in texture classification there exist various methods for feature extraction based on the use of filters and signal processing.

This paper is structured as follows. Literature survey is described in Sect. 2. Section 3 deals with the description of Shearlet Transform and the Sect. 4 deal with the feature extraction. Experimental study is explained in Sect. 5. The results are concluded in Sect. 6 and finally References.

2 Literature Survey

Texture analysis is broadly classified into three categories: Pixel based method, local feature based method and Region based method. Pixel based method uses gray level co-occurrence matrices, difference histogram and energy measurements and Local Binary Patterns (LBP). Local feature based method uses edges of local features and generalization of co-occurrence matrices. Region based method uses region growing and topographic models [4].

The spatial distributions of gray values are obtained by calculating the image features at each point and then developing a set of measurements from the distributions of the local texture features. Statistical approaches are the one which are use for natural images that are mostly irregular and in some cases they are repeated. It provides descriptions of textures as fine, rough, smooth, coarse, etc. Thus the experimental measures of texture are based on the original size, which could be the average area of the primitives of reasonably constant gray level [5].

Kaiser have explained about statistical method which examines the spatial distribution of gray values by computing the local features at each point in an image, and extracting a set of statistics from the distribution of local features [6]. This method is also used for examining the regularity and roughness of texture including autocorrelation function. Haralick et al. [7] proposed the equivalent procedure to identify the spatial uniformity of shapes called structural elements in a binary image. When these element themselves are single resolution cells, the autocorrelation function of the binary image is given. The size of co-occurrence matrix obtained will be same as that of the number of threshold levels. The most widely used geometric methods are the use of co-occurrence features and gray level differences which have stimulated a variety of modifications further. The process also includes the signed differences [8] and the LBP (Local Binary Pattern) operator [9]. Other statistical approaches which are already present are autocorrelation function that is used for examining the properties of the texture, and gray level run lengths. The above methods produce low classification efficiency.

Texture classification is an essential process for computer vision and image recognition applications. Texture classification is achieved by four methods such as structural, statistical, model-based and multiscale transform based methods. Multiscale methods are most widely used, since the multiresolution and directional representations of the transforms are possible with the human observation of the textured images. Some of the multiscale transforms that are most commonly used are the Gabor transform [10], the Wavelet transform [11], the Ridgelet transform [12], Curvelet transform and the Contourlet transform [13]. The results of this research work was compared with recent method results like BP-MD DONG et al. (2015), MCC-KNN Dong et al. and DST-ED Kanchana et al. (2013).

3 Shearlet Transform

Shearlets are the multiscale framework which allows to efficiently encode anisotropic features in multivariate problem classes [14]. Shearlets are considered as the sloping waveforms, with directions organized by the shear parameters, and they become gradually thin at adequate scales (for the value such as $p \to 0$). One of the most important properties of shearlet is the fact that they provide optimally sparse approximation for cartoon-like functions. The shearlet transform is not like the traditional wavelet transform which are only good at representing point singularities and do not have the ability to detect lines and curves. But the shearlet transform contains two parameters namely, the scaling parameter 'a' and the translation parameter 't' that are used for analyzing the directions [15]. The shearlet transform was developed next to contourlet to overcome the limitations present in wavelets and contourlets. By using shearlet transform, the image textures are represented as a simple but accurate mathematical framework which is also considered as the useful tool for the geometrical representation of multidimensional information, and this process is more usual for execution of the different applications.

The shearlet transform decomposes the input image into number of subband images containing the high rate of recurrence subband images and low rate of recurrence subband images. The magnitude of each and every shearlet subband has the same magnitude as that of the initial image [16]. The decomposition is highly redundant. In general, two different types of shearlet systems are utilized today: Band-limited shearlet systems and compactly supported shearlet systems. Regarding those from an algorithmic viewpoint, both have their particular advantages and disadvantages: Algorithmic realizations of the band limited shearlet transform on the one hand has the higher calculation difficulty. However, on the other hand, the process of handling the seismic data requires high localization [17]. The compactly supported shearlet transform are much faster and have the advantage of achieving a high accuracy in spatial domain.

The Fig. 1 represents the two level multiscale, shift invariant and multidirectional image decomposition. During first level decomposition it will give one approximation and n directional detailed bands. The first level approximation is given for further level decomposion. Though the Shearlet Transform consist of necessary properties of shift invariance, features were extracted from the texture images by using these properties.

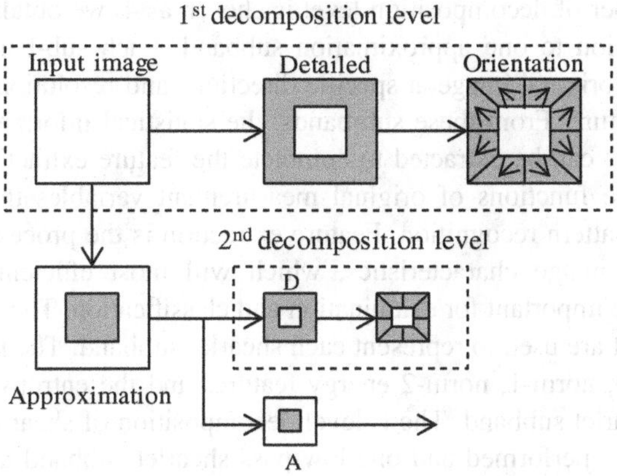

1st decomposition level

Input image   Detailed   Orientation

2nd decomposition level

D

Approximation

A

Fig. 1. Filter bank structure of Shearlet transform

4 Proposed System

In many texture classification system the image cannot be efficiently analyzed at various scales and directions. Therefore, an efficient way to obtain a multi-resolution and multi-direction representation of texture images based on shearlet transform is proposed. The proposed method for texture classification system based on shearlet transform is shown in Fig. 2. In general, a typical classification system mainly consists of two phases; feature extraction phase and classification phase.

Feature Extraction Phase. In this phase, the texture images are decomposed into two, namely the approximation (low frequency) component and the detailed (high frequency) components of various subbands by the Shearlet transform at different resolution levels.

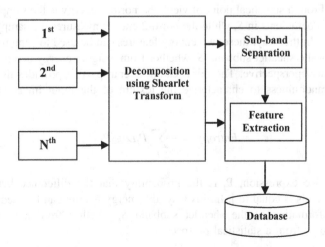

Fig. 2. Feature extraction stage of Shearlet transform

Since the number of decomposition level is chosen as 4, we obtain 16 subbands in each level in addition to one approximation subband. Each sub-band represents the components of the original image at specific directions and resolutions with the actual size of original texture. From these sub-bands, the statistical information which characterizes the texture can be extracted to complete the feature extraction.

Features are the functions of original measurement variables that are useful for classification and pattern recognition. Feature extraction is the process of describing a set of features or image characteristics, which will most efficiently represent the information that are important for examination and classification. The subband energies are the features that are used to represent each shearlet subband. The most widely-used energy features are, norm-1, norm-2 energy features and the entropy that are used to represent each shearlet subband. The L-level decomposition of shearlet transform on a given texture is first performed and one low-pass shearlet subband and M directional subbands at each scale were obtained. Then each shearlet subband are denoted as

$$S = \sum_{\tau=1}^{N} z_\tau$$

The norm-1 energy and norm-2 energy features are defined as

$$e1 = \frac{1}{N} \sum_{\tau=1}^{N} |z_\tau|$$

and,

$$e2 = \sqrt{\frac{1}{N} \sum_{\tau=1}^{N} |z_\tau|^2}$$

respectively. From a statistical point of view, the norm-1 energy is the sample mean of the moduli of coefficients in S, while the norm-2 energy measures the sample standard deviation [18]. In this way, these two energy features can be used to capture the visual information of the shearlet subband S, whether from a signal energy demonstration or from a statistical perspective. The third feature is entropy that provides the statistical measure of randomness to characterize the texture of the input image. Entropy is calculated as

$$Entropy = -\sum_{j} P_j Log_2 P_j$$

In the above expression, P_j is the probability that the difference between two neighboring pixels is equal to j. In this way, the energy feature can be used to capture the visual information of the shearlet subband S, whether from a signal energy demonstration or from a statistical perspective.

A. *Classification phase*

In the classification phase, the same kind of features are extracted and compared with the database obtained in the feature extraction stage. The features of all the subbands are fused together to form the feature vector of the related texture image. Similarly, the proposed features are extracted for all training texture samples and stored in the database for classification. Figure 3 show the classification stage of the proposed system. The nearest neighbor classifier and SVM classifier are designed to classify the unknown texture image into known texture class.

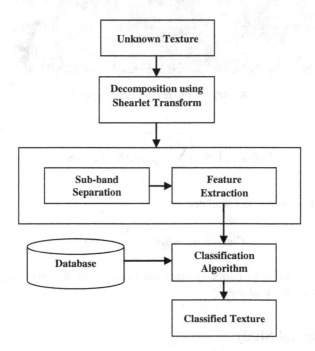

Fig. 3. Classification stage of Shearlet Transform

The texture image to be classified is decomposed by means of Shearlet Transform and the feature vector is extracted as in the training phase. It should be noted that the number of decomposition level, number of directions and block size should be similar in both the phases to avoid failure of classifier. The classification is done by using minimum distance measure and SVM classification. The City Block distance measure is used in the proposed method. The performance measure of the proposed texture classification system is the classification accuracy which is measured as the percentage of test images classified into the exact texture. The classification rate is based on the number of correctly classified test textures (Fig. 4).

The Minimum Distance between the test image to training image can be calculated by using the following equation:

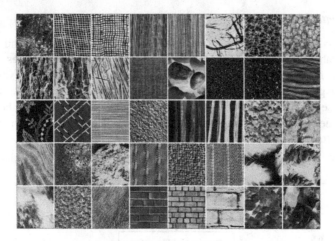

Fig. 4. Brodatz images used in experiments

$$D(f_{test}, f_{lib}) = \sqrt{\sum_{i=0}^{N} \left(f_{(test)i}(x) - f_{(lib)i}(m)\right)^2}$$

The classification rate is calculated as

$$Classification\,Rate = \frac{M}{N} \times 100$$

where M is the number of correctly classified images and N is the total number of images present in the database.

5   Experimental Study

In this section, the performance of the proposed texture classification algorithm based on Shearlet Transform is described. Brodatz texture images are used to estimate the performance of the proposed system. The size of the Brodatz texture images used are 512×512 and 640×640 pixels and the images are gray scale images. The databases consist of different classes which are divided based on the basic character of the pictures. The Brodatz database [19] consisting of 112 texture images of different class. Every image is divided into multiple patches from which some patches are used as training samples and others as tesing samples. The Shearlet Transform features are obtained and stored. The textures were classified by calculating the distance using Minimum distance classifier and SVM classifier. The classification results were obtained using the two classifiers and compared for Brodatz album. The classification rate is based on the number of correctly classified test textures.

The experiments were conducted with 12 basic textures of Brodatz dataset and with 40 texture images of Brodatz dataset. Each texture image is sub divided into sixteen sub images and features such as mean, standard deviation and entropy were used for classification.

In Table 1, *C1* and *C2* represents the experiment with 16 sub images for training and testing using minimum distance classifier and SVM classifier. *C3* and *C4* represents the experiment with 12 sub images for training and 4 sub images for testing using minimum distance classifier and SVM classifier. *C5* and *C6* represents the experiment with 8 sub images for training and 8 sub images for testing using minimum distance classifier and SVM classifier.

Table 1. Results for texture classification using shearlet transform

Images	Correct classification (%)					
	C1	C2	C3	C4	C5	C6
Bark	100	100	100	100	75	100
Brick	93.7	100	75	100	87.5	100
Bubbles	100	100	100	100	75	100
Grass	100	100	100	100	62.5	75
Leather	100	100	100	100	100	100
Pigskin	93.7	100	100	100	100	100
Sand	100	93.75	100	100	100	87.5
Straw	100	100	100	100	62.5	100
Weave	87.5	100	75	75	75	50
Water	100	100	100	100	62.5	75
Wool	100	93.75	100	100	87.5	100
Wood	100	100	100	100	75	100
Mean classification rate(%)	97.9	98.95	95.8	97.9	80.2	90.6

In Table 2. For 40 Brodatz textures F1 represents the experiment with 16 sub images for training testing, F2 represents the experiment with 12 sub images for training and 4 sub images for testing and F3 represents the experiment with 8 sub images for training and 8 sub images for testing (Fig. 5).

Table 2. Results for texture classification using shearlet transform energy features for 40 textures

CLASSIFIERS	F1	F2	F3
MD	94.5%	93.65%	86.45%
SVM	97.5%	95.35%	94.7%

1 – Experiment with 16 sub images for training and testing for all 40 texture images.
2 – Experiment with 12 sub images for training and 4 sub images for testing for all 40 texture images.
3 – Experiment with 8 sub images for training and 8 sub images for testing for all 40 texture images.

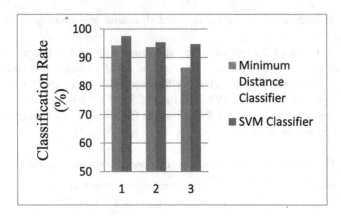

Fig. 5. Comparison result for minimum distance and SVM classifier using 40 Brodatz textures

In Table 3. the comparative analysis of the proposed system with other techniques in the literature in terms of classification accuracy is provided. It is observed that the proposed system outperforms all other methods in terms of average classification accuracy.

Table 3. Comparative analysis of the proposed system with the other techniques in the literature.

Methods	Classification accuracy (%)
BP-MD Dong et al. (2015)	80.58
MCC-KNN Dong et al. (2015)	72.25
DST-ED Kanchana et al. (2013)	94.24
Proposed method	97.5

6   Conclusion

From the experimental analysis it is inferred that the proposed feature set produces good classification rate. This method is done with natural Brodatz texture images. The success rate obtained for Shearlet Transform is improved when compared with other transforms. In order to overcome the inherent complexity of the Shearlet coefficients, we extract energy features such as mean, standard deviation and entropy from the Shearlet subbands and model their dependences using the Minimum Distance Classifier and the SVM classifier. It is proved that Shearlets exhibit high directional sensitivity. The high directional sensitivity of the Shearlet transform and its optimal approximation properties will lead to the enhancement of many image processing applications. Thus overcoming the disadvantages of the Contourlet transform, that there are no restrictions on the number of directions for the shearing. It is inferred that the proposed feature set produces good classification rate for SVM classifier.

References

1. Liu, L., Fieguth, P.W.: Texture classification from random features. IEEE Trans. Pattern Anal. Mach. Intell. **34**(3), 574–586 (2012)
2. Raghu, P.P., Poongodi, R., Yegnanarayana, B.: Unsupervised texture classification using vector quantization and deterministic relaxation neural network. IEEE Trans. Image Process. **6**(10), 1376–1387 (1997)
3. Salem, Y.B., Nasri, S.: Texture classification of woven fabric based on a GLCM method and using multiclass support vector machine. In: SSD, pp. 1–8 (2009)
4. Dixit, A., Hedge, N.P.: Image texture analysis techniques-survey. In: 2013 Third International Conference on Advanced Computing & Communication Technologies, ©2013. IEEE (2013)
5. Yo-Ping, H., Tsun-Wei, C.: Retrieving interesting images using fuzzy image segmentation and fuzzy data mining model. In: Proceedings of IEEE Annual Meeting of the Fuzzy Information, vol. 2, pp. 623–628, June 2004
6. Srinivasan, G.N., Shobha, G.: Statistical texture analysis. In: Proceedings of World Academy of Science, Engineering and Technology, vol. 36, December 2008. ISSN 2070-3740
7. Haralick, R.M., Shanmugan, K., Dinstein, I.: Texture features for image classification. IEEE Trans. Syst. Man Cybern. **3**(6), 610–621 (1973)
8. Kaizer, H.: A Quantification of Texture on Aerial Photographs. Boston University Technical Laboratory, Boston (1995)
9. Pitikainen, M., Ojala, T., Maenpaa, T.: Multiresolution gray-scale and rotation invariant texture classification with local binary patterns. PAMI **24**(7), 971–987 (2002)
10. Liao, S., Law, M.W.K., Chung, A.C.S.: Dominant local binary patterns for texture classification. IEEE Trans. Image Process. **18**(5), 1107–1118 (2009)
11. Unser, M.: Texture classification and segmentaion using wavelet frames. IEEE Trans. Image Process. **4**(11), 1549–1560 (1995)
12. Arivazhagan, S., Ganesan, L., Subash Kumar, T.G.: Texture classification using ridgelet transform. Pattern Recogn. Lett. **27**(16), 1875–1883 (2006)
13. Po, D.D.-Y., Do, M.N.: Directional multiscale modeling of images using the contourlet transform. IEEE Trans. Image Process. **15**(6), 1610–1620 (2006)
14. Guo, K., Labate, D.: Optimally sparse multidimensional representation using shearlets. SIAM J. Math. Anal. **39**(1), 298–318 (2007)
15. Robson, W., Ricardo, S., da Silva, D., Davis, L.S., Pedrini, H.: A novel feature descriptor based on the shearlet transform. In: 2011 18th IEEE International Conference on Image Processing (2011)
16. Dong, Y., Tao, D., Li, X., Ma, J., Pu, J.: Texture classification and retrieval using shearlets and linear regression. IEEE Trans. Cybern. **45**(3), 358–369 (2015)
17. Easley, G., Labate, D., Lim, W.-Q.: Sparse directional image representations using the discrete shearlet transform. Appl. Comput. Harmon. Anal. **25**(1), 25–46 (2008)
18. Kanchana, M., Varalakshmi, P.: Texture classification using discrete shearlet transform. Int. J. Sci. Res. June 2013. ISSN No 2277-8179
19. The Brodatz Texture database with 112 images. http://www.ux.uis.no/tranden/brodatz.html

Enhanced ℓ – Diversity Algorithm for Privacy Preserving Data Mining

R. Praveena Priyadarsini$^{(\boxtimes)}$, S. Sivakumari, and P. Amudha

Department of Computer Science and Engineering, Avinashilingam Institute
for Home Science and Higher Education for Women, Coimbatore, India
praveena.priya04@gmail.com,
prof.sivakumari@gmail.com, amudharul@gmail.com

Abstract. With the increase in use of e-technologies, large amount of digital data are available on-line. These data are used by both internal and external sources for analysis and research. This digital data contain sensitive and personal information about the entities on which the data are collected. Due to this sensitive nature of such information, it needs some privacy preservation procedure to be applied before releasing the data to third parties. The privacy preservation should be applied on the data such that its utility during data mining does not get reduced. ℓ-Diversity is an anonymization algorithm that can be applied on dataset with one sensitive attribute. Real life data contain numerous sensitive attributes that have to be privacy preserved before publishing it for research. This paper proposes an Enhanced ℓ-diversity algorithm that can diversify multiple sensitive attributes without partitioning the dataset. Two datasets namely, bench mark Adult dataset and Real life Medical dataset are used for experimentation in this work. The privacy preserved datasets using the proposed algorithm are compared for its utility with ℓ-diversified dataset for single sensitive attribute and original dataset. The results show that the proposed algorithm privacy preserved datasets have good utility on selected classification algorithms taken for study.

Keywords: ℓ-diversity · Enhanced ℓ-diversity · Multiple sensitive attributes · Privacy preservation

1 Introduction

Privacy-preserving data mining refers to the area of data mining that seeks to safeguard sensitive information from unofficial disclosure [14]. Government agencies and other organization frequently require to publish their data for investigate and other purposes. These micro-data contain records about individuals and organizations which contain sensitive and personal information about a person, a household, or society. Thus privacy preservation is needed to hide these sensitive information before they are used for research and publishing. The attributes in the micro- data can be divided into three types namely, sensitive attributes, non-sensitive attributes and Quasi-Identifiers from privacy preservation point of view. Sensitive attributes contain personal privacy information that must identify the exact information of individuals. Quasi-identifiers

© Springer Nature Singapore Pte Ltd. 2016
S. Subramanian et al. (Eds.): CSI 2016, CCIS 679, pp. 14–23, 2016.
DOI: 10.1007/978-981-10-3274-5_2

are the set of attributes that can be linked with other datasets which is publically available to identify individual's private data. There are many techniques used for privacy preservation. The most commonly used technique for privacy preservation is anonymization technique. Here, the sensitive information is replaced by some random value based on probability distribution.

ℓ-Diversity proposed by Machanavajjhala et al. [1] is an extension of the k-anonymity model. This method overcomes homogeneous and background knowledge attack from k-anonymity technique. This requires every group of attributes should contain ℓ-diversified value to have ℓ-distinct values within the bucket. The major drawback of algorithm is that, it cannot handle multiple sensitive attributes. Most of the extensions in this algorithm have partitioned the dataset to help ℓ-diversity algorithm to accommodate multiple sensitive attributes. The proposed work proposes enhanced ℓ–diversity algorithm that can handle multiple sensitive attributes without partitioning the dataset. The proposed technique applies on three sensitive attributes within the dataset and the privacy preserved datasets from the proposed algorithm is evaluated based on its utility on classification function of data mining.

The paper is organized as follows: Sect. 1 gives the introduction. Section 2 presents literature survey. Section 3 gives the dataset description; Sect. 4 discusses about the proposed Enhanced ℓ-diversity algorithm. Section 5 presents experimental results and discussions and Sect. 6 gives the conclusions.

2 Literature Survey

Aggarwal et al. discussed about techniques like k-anonymity, ℓ-diversity, randomization, condensation and sanitization methods to achieve the privacy of sensitive data in a dataset [14]. Fung et al. discussed about techniques used to preserve the sensitive information before publishing [11].

Aggarwal and Yu [4] proposed a condensation approach for privacy preservation such that dataset are anonymized to a data that closely matches the characteristics of the original data and preserves the correlation among the attribute. Han et al. [5] proposed two anonymization methods to preserve privacy in mixed data. Both the methods MEGA and TSCKA affectively preserve the privacy of mixed data and also have good data quality. Han et al. proposed method called SLOMS where two sensitive attributes that have high chi-square correlation between them are clustered into separate tables. The quasi identifiers of these sensitive attributes are generalized [8].

Xiao and Tao [17] has proposed a method called Anatomy that releases the sensitive and quasi attributes in the dataset as a separate table. A grouping mechanism is performed that preserves privacy as well as captures correlation among the attributes in the dataset.

Machanavajjhala has discussed about the technique known as ℓ-diversity which overcomes the drawbacks of k-anonymity technique and provides highly efficient privacy preservation [1]. ℓ–Diversity requires the equivalence class to have at least ℓ well represented values for each sensitive attribute. But the most important limitation of ℓ-diversity is that it is not sufficient to prevent attribute disclosure attack. Also, when sensitive values of the group are semantically close there is a breach in privacy on a

ℓ-diversified dataset. This limitation of ℓ-diversity is addressed by t–closeness which defines that an equivalence class is t–close if the distance between the distribution of a sensitive attribute in this class and distribution of the attribute in the whole table is no more than a threshold 't' [1]. David Kifer and Johannes Gehrka introduced anonymized marginal to increase the utility of the k–anonymous and ℓ–diverse tables [2]. Zak-erzadeh et al. proposed a generalized approach that uses inter-attribute correlates to decompose the data into vertical dataset. An anonymization process that complies with k–anonymity, ℓ- diversity and t–closeness models has been applied on the partitions for privacy preservation [3]. Sattar et al. [6] proposed a probabilistic d-linkable model to mitigate composition attack. This model uses partition based k–anonymization approach. Huiwang and Ruilin Liu proposed (d,l) inference model to anonymization micro-data with unsafe functional dependency with low information loss [7]. Das and Bhathacharrya proposed decomposition algorithm which does ℓ-diversity on a multiple sensitive attributes using partitioning method [9]. Aggarwal, discussed about main-taining ℓ-diversity across "r" sensitive attributes [10].

Grigorios Loukides et al. introduced a rule-based privacy model that allows data publishers to express fine-grained protection requirements for both identity and sen-sitive information disclosure using two different two different anonymization algo-rithms [12]. First algorithm is to recursively generalize data with low information loss. Second algorithm greatly improved scalability while maintaining low information loss.

Tamas S. Gal and Zhiyuan Chen proposed a technique which treats all the sensitive attributes uniformly. This method allows different degrees of diversity on different attributes at every horizontal partitioning of the data. This horizontal partitions are then anatomized separately [13].

3 Dataset Descriptions

Real time Medical Dataset and Adult Dataset from UCI Knowledge Discovery Archive database [15] are used for experimentation in this work. The adult dataset contains 15 attributes including one class attribute which has two categorical values, ">50 K" and "≤50 K". The non-class attributes are age, workclass, fnlwgt, education, education-num, marital-status, occupation, relationship, race, sex, capital-gain, capital-loss, hours-per-week, and native-country. The description of Adult dataset is given in Table 1.

The Medical dataset consists of 150 records which contains patient details. There are 13 attributes including the one class attribute with categorical values "Breathing" and "Non-Breathing". The non-class attributes are unique id, Age, Sex, Address,

Table 1. Adult dataset

Dataset	Description
Attribute characteristics	Categorical, integer
Number of instances	48842
Number of attributes	14
Missing values	Yes
No. of classes	2

Table 2. Medical dataset

Dataset	Description
Attribute characteristics:	Categorical, integer
Number of instances:	150
Number of attributes:	13
Missing values	No
No. of classes	2

Occupation, Complaints, Appetite, Mental generals, Built, Height, Weight, and Diagnosis. The description of Real life Medical dataset is given in Table 2.

4 Enhanced ℓ-Diversity Algorithm

The attributes in a dataset can divided in to three types namely, sensitive attributes, non-sensitive attributes and Quasi-Identifiers from privacy preservation point of view. Sensitive attributes are those that contain personal privacy information to identify the exact information of individuals. Quasi-identifiers are the set of attributes that can be linked with other datasets which is publically available to identify individual's private data. ℓ-Diversity technique is an extension of the k-anonymity model where a given record maps onto at least k other records in the data and results in loss of data. The ℓ-diversity requires every group of attributes should contain ℓ-diversified value to have ℓ-distinct values within the bucket. The major drawback of ℓ-diversity algorithm is that it cannot handle multiple sensitive attributes.

Tamas S. Gal and Zhiyuan Chen proposed an algorithm which partitions the data and applies anatomy to accommodate multiple sensitive attributes [13]. In this work the proposed algorithm tries to accommodate multiple sensitive attributes for ℓ-diversity by applying few conditions on setting the bucket size and accommodating the various attribute values of sensitive attributes within this bucket. In order to minimize the loss of information, if the ℓ-value for diversification is much lesser than the values present in the Sensitive attribute, generalization is applied. For categorical sensitive attributes where each SA has V_n different values, the ℓ value for diversity is set based on the occurrence of distinct values in QI attributes. The number of distinct values in the quasi identifiers of the dataset are recorded and compared with distinct value of SA attributes. The ℓ-diversity is applied for 1 attribute, 2 attributes and 3 attributes.

Let $\{q_1, q_2, \ldots q_n \in D\}$ be the set of QI attributes of the dataset D. The distinct values held by these attributes are $\{V_1, V_2, \ldots .. V_N \in D\}$. Let $\{SA_1, SA_2 \ldots .SA_n \in D\}$ be the set of Sensitive Attributes (SA) in dataset D and the distinct value held by these SA be $\{Y_1, Y_2, \ldots Y_n \in SA\}$.

Let the minimum $(V_n) = A$, maximum $(V_n) = B$, minimum $(Y_n) = C$ and maximum $(Y_n) = D$

The ℓ value and bucket size during diversification is set based on the following rules given in Eqs. (1, 2 and 3):

If A $=$ C then set bucket size (Z) and ℓ − value of the dataset as C   (1)

If C $>$ A then set the values Z $=$ L $=$ A   (2)

If C $<$ A then set the values Z $=$ L $=$ C   (3)

Also, to decrease the information loss in SA attributes when Z value is set as min (V_n) generalization is applied on SA_n based on Eq. (4).

If $Y_n((SA_n)) > 2 \times A$) then generalize Y_n, $SA_n \rightarrow A$ (4)

Thus, each sensitive attributes in D will be min (V_n) diverse or min (Y_n) diverse.

For example, in the proposed work the QI attributes in the adult dataset have the distinct values $\{V_1, V_2, \ldots \ldots V_4\}$ as 7, 10, 13, 6. The distinct values of two SA attributes $\{Y_1, Y_n\}$ in the data set be 13, 6. Then, min $(V_n) = 6$ and min $(Y_n) = 6$.

Hence rule 1 applies and bucket size Z and ℓ- value for diversity for the dataset is set as 6. Since the SA attribute set has 13 distinct values inside it which is greater than two times min (V_n), then generalization is applied to reduce to min (V_n). Thus each bucket will be min (V_n) anonymized and all the equivalence class will have a well-defined min (V_n) diverse values.

As Y_n (SA) $>$ two times min (V_n) then sensitive attributes are generalized to min (V_n) value. Thus, each sensitive attributes is ℓ-diverse that helps in preventing membership disclosure attack.

```
Let {q1, q2....... qₙ∈ D }, QI ⊆ D
And { SA₁,SA₂........SAₙ∈ D}, SA ⊆D
SAₙ ∩ QIₙ = φ
Let {V₁,V₂,....Vₙ ∈ q} be the distant values of QI set
{y₁,y₂....yₙ ∈ SA} be distinct values of SA set
Bucket size (Z) and diversity value L is
If    min(yₙ) = min(vₙ)    then   Z = L = min (vₙ)
Else  If    min(yₙ) > min(vₙ)    then   Z = L = min (vₙ)
Else    If   min(yₙ) < min(vₙ)    then   Z = L = min (yₙ)
For SAᵢ where i = 1 to n
If    SAᵢ(Yₙ) > 2 × min(vₙ)      then
Generalized        SAᵢ(Yₙ) → min(Vₙ) diverse
    ℓ-diversity[SA] ;       k-anonymity[QI] ;
Perturbed Dataset = D̄= QI ∪ SA ∪ NSA
```

Fig. 1. Enhanced ℓ-diversity algorithm

The proposed algorithm is given in Fig. 1. Using the proposed algorithm, the privacy preserved dataset will be having min (V_n) diversity or min (Y_n) on all SA attributes of the dataset. Also all the quasi attributes will also be min (Y_n) or min (V_n) anonymized. Thus the proposed algorithm minimizes attribute value deletion during ℓ - diversity using generalization, and increases the utility of the privacy preserved dataset.

In Adult dataset, attributes described in Table 1, "Occupation", "Relationship" and "Work class" are set as sensitive attributes. Suppose if third party knows individual's occupation details then, he/she may easily identify other details about a person which violates their privacy. In the real time medical data set "Complaints", "Diagnosis" and "Mental Generals" attributes are set as sensitive attributes. The attributes is set as sensitive based on Information gain value of the attribute on class attribute.

The privacy preserved datasets by applying the algorithm incrementally on more than one SA attributes are compared on its utility on classification algorithm using Weka simulator. The level privacy increases in the privacy preserved versions of datasets as the number of attributes subjected to ℓ - diversification increases. These versions can be distributed to users with different trust levels.

5   Experimental Results and Discussion

The two datasets taken for experimentation in this work are: Adult dataset and real world Medical dataset. The ℓ-diversity algorithm is coded using Java as front-end and MySQL as back-end. The various dataset obtained after applying the enhanced ℓ-diversity algorithm on both the datasets are given in Table 1.

Table 3. Privacy preserved adult and medical dataset versions

S.No	Privacy preserved Datasets	Abbreviations
1	ℓ-Diversity with 1 SA for adult dataset	L-D-1-SA- adult dataset
2	ℓ-Diversity with 2 SA for adult dataset	L-D-2-SA- adult dataset
3	ℓ-Diversity with 3 SA for adult dataset	L-D-3-SA -adult dataset
4	ℓ-Diversity with 1 SA for medical dataset	L-D-1-SA- medical dataset
5	ℓ-Diversity with 2 SA for medical dataset	L-D-2-SA- medical dataset
6	ℓ-Diversity with 3 SA for medical dataset	L-D-3-SA- medical dataset

These datasets are compared on their utility using classification accuracy and Receiver Operating Characteristic (ROC) metrics on classification algorithms like Naive Bayes, C4.5 and Ripper using Weka simulator [16].

The equation for calculating classification accuracy is given in Eq. (5),

$$\text{Classification accuracy} = \frac{\textit{No. of correctly classified tuples}}{\textit{Total No. of tuples in the dataset}} \tag{5}$$

ROC-Area returns the probability or ranking for the predicted class for each tuple present in the dataset. It is a plot of the true positive rate against the false positive rate. The Roc-Area values range 0 to 1.0. The closer the value of area to 0.5 the less accurate and value will be 1.0 for the perfect accuracy [18].

The classification accuracy of the three versions of privacy preserved adult and Medical datasets is measured using classification accuracy of three classifiers namely Naïve Bayes, C4.5 and Ripper algorithms are given in Figs. 2 and 3.

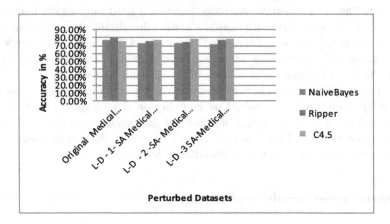

Fig. 2. Comparison of classification accuracy of privacy preserved adult datasets

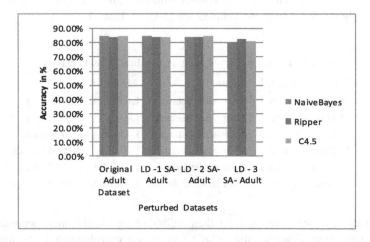

Fig. 3. Comparison of Classification Accuracy of Privacy Preserved Medical Datasets

Figure 2 shows that on all the privacy preserved datasets, Naïve Bayes classifier has a decrease of 4% in accuracy. On Ripper algorithm there is a decrease in 2% accuracy and on C4.5 algorithm there is a decrease in 1% accuracy. C4.5 classifiers have good accuracy when compared with all other classifiers on all the three privacy preserved datasets. When ℓ-diversity is extended to more number of SA, there is loss of utility for all other classifiers.

Figure 3 shows that the utility of ℓ-diversified medical dataset on multiple sensitive attributes is compared on all the three classifiers. The results show that in Naïve Bayes classifier there is decrease of 4% accuracy. On Ripper algorithm there is a decrease in 2% accuracy and on C4.5 algorithm there is an increase in 2% accuracy. C4.5 classifiers have good accuracy when compared with all other classifiers for all the privacy preserved datasets. When ℓ-diversity is extended to more number of SA then there is a loss of utility (Fig. 3).

The ROC measurements of the privacy preserved versions of adult and medical datasets are tabulated in Tables 4 and 5.

Table 4. Comparing the ROC values of privacy preserved adult datasets

Dataset	ROC area for Naïve Bayes	ROC area for Ripper	ROC area for C4.5
OAD	0.892	0.799	0.791
ℓ-Diversity - 1 SA -adult dataset	0.888	0.796	0.762
ℓ-Diversity - 2 SA -adult dataset	0.881	0.776	0.764
ℓ-Diversity -3 SA - adult dataset	0.656	0.58	0.55

Table 5. Comparing the ROC values of privacy preserved medical dataset

Dataset	ROC area for Naive Bayes	ROC area for Ripper	ROC area for C4.5
OMD	0.78	0.634	0.47
ℓ-Diversity -1 SA –medical dataset	0.428	0.454	0.468
ℓ-Diversity - 2 SA- medical dataset	0.356	0.464	0.468
ℓ-Diversity - 3 SA- medical dataset	0.343	0.51	0.468

ROC values on Adult ℓ-diversified datasets in Table 4 shows that ℓ-Diversity - 3 SA – Adult dataset has the least ROC values when compared to other L-diversified and original dataset. On medical dataset all the privacy preserved L-diversified datasets have lesser ROC values on Naïve Bayes and Ripper algorithm. ROC value of C4.5 algorithm remains the same as original dataset on all the privacy preserved versions on medical datasets.

Thus, the results indicate that there is a very little decrease in accuracy on the Enhanced L-diversified versions of privacy preserved datasets on both the datasets taken for study. Thus, Enhanced ℓ-diversity algorithm can produce anonymized versions of privacy preserved dataset for multiple SA attributes without vertically partitioning the datasets. Since all the Quasi Identifiers are k-anonymized and the SA attributes are L-diversified the proposed algorithm privacy preserved datasets can prevent homogeneous and background knowledge.

6 Conclusions

Real life datasets and high dimensional datasets contain number of sensitive attributes. In order to apply ℓ-diversity on these types of datasets which contain more than one sensitive attributes, this work proposes Enhanced ℓ-diversity algorithm. The proposed algorithm can diversify more than one sensitive attribute in a dataset without partitioning the dataset. Thus, the proposed algorithm tries to eliminate the drawback of ℓ-diversity algorithm by extending it to accommodate multiple attributes. When the proposed privacy preserved ℓ-diversified datasets are evaluated for their utility on selected classification algorithms they exhibit very less decrease in utility when compared with traditional ℓ-diversity algorithm and original dataset. Thus the proposed algorithm can produce ℓ-diversity on datasets with multiple sensitive attributes with negligible decrease in utility. As future extension feature reduction techniques can be used along with this method to perform privacy preservation on big data.

References

1. Machanavajjhala, A., Kifer, D., Gehrke, J., Venkitasubramaniam, M.: l-diversity: privacy beyond k-anonymity. ACM Trans. Knowl. Discov. Data (TKDD) **1**(1), 3 (2007)
2. Kifer, D., Gehrke, J.: Injecting utility into anonymized datasets. In: ACM SIGMOD International Conference on Management of Data, pp. 217–228 (2006)
3. Zakerzadeh, H., Aggrawal, C.C., Barker, K.: Towards breaking the curse of dimensionality for high-dimensional privacy: An Extended Version (2014). arXiv preprint arXiv:1401.1174
4. Aggarwal, C.C., Yu, P.S.: A condensation approach to privacy preserving data mining. In: Bertino, E., Christodoulakis, S., Plexousakis, D., Christophides, V., Koubarakis, M., Böhm, K., Ferrari, E. (eds.) EDBT 2004. LNCS, vol. 2992, pp. 183–199. Springer, Heidelberg (2004). doi:10.1007/978-3-540-24741-8_12
5. Han, J., Yu, J., Mo, Y., Lu, J., Liu, H.: MAGE: A semantics retaining K-anonymization method for mixed data. Knowl.-Based Syst. **55**, 75–86 (2014)
6. Sattar, A.S., Li, J., Liu, J., Heatherly, R., Malin, B.: A probabilistic approach to mitigate composition attacks on privacy in non-coordinated environments. Knowl.-Based Syst. **67**, 361–372 (2014)
7. Wang, H., Liu, R.: Privacy-preserving publishing microdata with full functional dependencies. Data Knowl. Eng. **70**, 3 (2011)
8. Han, J., Luo, F., Lu, J., Peng, H.: SLOMS: A privacy preserving data publishing method for multiple sensitive attributes microdata. J. Softw. **8**(12), 3096–3104 (2013)
9. Das, D., Bhattacharyya, D.K.: Decomposition +: Improving ℓ-Diversity for Multiple Sensitive Attributes. In: International Conference on Computer Science and Information Technology, pp. 403–412. Springer, Heidelberg (2012)
10. Aggarwal, C.C.: Privacy and the dimensionality curse. In: Aggarwal, C.C., Yu, P.S. (eds.) Privacy-Preserving Data Mining, pp. 433–460. Springer US, New York (2008)
11. Fung, B., Wang, K., Chen, R., Yu, P.S.: Privacy-preserving data publishing: a survey of recent developments. ACM Comput. Surv. **42**(4), 14 (2010)
12. Loukides, G., Gkoulalas-Divanis, A., Shao, J.: Efficient and flexible anonymization of transaction data. Knowl. Inf. Syst. **36**(1), 153–210 (2013)

13. Gal, T.S., Chen, Z., Gangopadhyay, A.: A privacy protection model for patient data with multiple sensitive attributes. IGI Global, pp. 28–44 (2008)
14. Aggarwal, C.C., Philip, S.Y.: A general survey of privacy-preserving data mining models and algorithms. In: Aggarwal, C.C., Yu, P.S. (eds.) privacy-preserving data mining, pp. 11–52. Springer, US, New York (2008)
15. http://archive.ics.uci.edu/ml
16. Hall, M., Frank, E., Holmes, G., Pfahringer, B., Reutemann, P., Witten, I.H.: The WEKA data mining software:an update. ACM SIGKDD Explor. Newsl. 11(1), 10–18 (2009)
17. Xiao, X., Tao, Y.: Anatomy: simple and effective privacy preservation. In: 32nd International Conference on Very Large Data Bases, pp. 139–150. VLDB Endowment (2006)
18. Han, J., Pei, J., Kamber, M.: Data Mining: Concepts and Techniques. Elsevier, Amsterdam (2011)

An Enhanced and Efficient Algorithm for Faster, Better and Accurate Edge Detection

S.N. Abhishek[1], Shriram K. Vasudevan[1(✉)], and R.M.D. Sundaram[2]

[1] Department of Computer Science and Engineering, Amrita School of
Engineering, Amrita Vishwa Vidyapeetham (University), Coimbatore, India
abshakesn@gmail.com, kv_shriram@cb.amrita.edu
[2] Software Architect, Wipro Technologies, Bangalore, India
sundaram.ramanathan@wipro.com

Abstract. Pictures are considered to be the capture of a real world scenario and we second that "pictures speak more than words". It is also understandable that information can also be taken out of an image. To extract the information out of the image, the most important feature is the edge. There are many techniques to detect edges. This paper has a detailed study about these different techniques and proposed an intelligent algorithm. Also an analysis of these algorithms on various datasets are done.

Keywords: Edge detection · Hybrid edge detetion · Hybrid Sobel · Hybrid Robert · Hybrid Prewitt · Intelligent edge detection

1 Introduction

An image is a two-dimensional picture that depicts a physical environment or object or person. It contains some information about the subject present. The most important part of the image is the edges. It determines the quality and clarity of the image. Good edges can illuminate the information present in the image easily. Also poor edges push the information into dark. In a nutshell, importance of an image is shown by its edges. Information present in an image can be extracted in many ways. But it requires the edges to be legible to extract the information successfully. The article deals with some famous techniques performed for edge detection. To increase the clarity in case of bad edges, some additional techniques like smoothing and filtering are used and the same have been discussed. Once the edges are filtered out of the image, they must be enhanced to get a clear idea about the image. This paper also throws light on the famous edge enhancement technique. This enhanced result can be used for further analysis and research. But it is primarily important to know what an image is and edge technically.

Edge is precisely referred to as collection of points in a digital image where there is a sharp change in image brightness or has discontinuities in brightness. Edge is also referred as:

- Points at which the gradient in one direction is high, whereas the orthogonal direction gradient is low.
- Set of points in a same border where an unexpected change in the colour intensity occurs.

S. Subramanian et al. (Eds.): CSI 2016, CCIS 679, pp. 24–41, 2016.
DOI: 10.1007/978-981-10-3274-5_3

Based on an extensive literature survey conducted it has been theoretically proved in many cases that Canny edge detector gives the best result compared to any other operator [5, 6]. So an interesting idea has been proposed to enhance some common algorithms using the canny edge detection kernel. The output of the original and enhanced operators are compared and the best result is given as output.

2 Literature Survey

Shriram et al. have done an extensive research on usage of an edge detection technique for automotive image processing and it been observed that canny edge detector sounds to be the best of all. Also the authors have concluded that canny edge detector gives good results with using a Mahindra Scorpio SUV as an example [1].

Lijun Ding et al. had done research to find out the existing flaws in canny edge detector and they have identified that some edges are missing. The authors have suggested some minute changes in the existing algorithm and they found it satisfactory [2].

James Mathews has explained why exactly someone needs an edge and he also introduces the basics of edge detection with some simple examples. Then the author has taken Sobel edge detector as the case and explained the way it works. The author also confirms that sobel edge detector can pick even the smaller details from the image [3].

Raman Maini has made a comparison of all existing algorithms for image detection techniques in matlab and the paper talks about the architecture of the algorithms in detail. They claim that techniques such as Canny, LoG (Laplacian of Gaussian), Robert, Prewitt, all perform well and they say Canny performs better than other algorithms when the input image has noise [4].

Lary Davis has made a detailed survey on the various edge detection techniques. They have analysed all the popular algorithms in detail and had given an extensive study. This as an old and fully packed paper in the field. This is very popular to serve budding researchers in the field and has more than 900 citations [5].

Tamar Peli et al. has evaluated the various edge detection operators in terms of performance measure. They had studied the performance of Robert, Hale and Rosenfeld operators in detail. They had taken the output in binary form and compared their magnitudes alone. They had also explained the various classification of the edge detection techniques [6].

Mamta Juneja et al. had evaluated the various edge detection techniques based on their performance. They have compared the edge maps among pairs of the algorithms like Laplacian and Sobel. The detailed t-test of the combinations has been studied to find the best technique. After studying the edge maps through statistical evaluation it has been stated that canny edge detector performs better [7].

3 Edge Detection

Edge detection is a set of mathematical method used to detect the edge of an image. Edges are formally referred as discontinuities. The same process of detecting discontinuities in one-dimensional signals is referred as step detection or change detection.

It is a fundamental tool for detecting features and extracting them out of an image. Edges extracted from images that are not trivial are affected by fragmentation or discontinuity. It may miss some edge segments or also include some false edges. Thus complicating interpretation. A pre-processing step for edge detection is noise removal. Noise means "unwanted signal". It is an unnecessary and irrelevant information present in an image. Filters are used for the purpose, especially Gaussian filter.

There are three major criteria for edge detection:

- Detection should have low error rates. Low error rates means accurately catching maximum possible edges present in the image.
- The edge point detected must be present accurately in the centre of the edge.
- Every edge should be marked only once. False edges should not be created due to Image noise [3, 6].

The edge detection techniques are broadly classified as:

- Search based
- Zero-crossing based

4 Canny

Canny edge detection algorithm is one of the famous and efficient algorithm used for edge detection. It was developed in 1986 by, the scientist of computational theory, John F Canny. It involves a step by step procedure of removing the unwanted information and noise from the image and isolates the important features and information. Then the intensity gradient of the image is found and further optimized to obtain the edges of the image. It works only on images of PGM (Portable GrayMap) and PPM (Portable PixMap) format. It is being executed in a series of 5 steps:

4.1 Smoothing

It is done by removing the noise out of the image. This is done by performing Gaussian filter over the image. The filter convolves with the image. The equation of the Gaussian filter kernel is given in Eq. 1.

$$H_{ij} = \frac{1}{2\pi\sigma^2} \exp\left(-\frac{(i-k-1)^2 + (j-k-1)^2}{2\sigma^2}\right) \tag{1}$$

Here 'i' and 'j' are the index numbers of an entry in the matrix. 'σ' is the standard deviation of the distribution. Whereas 'k' comes from the size of the kernel. The kernel thus formed using the formula is convoluted with the image as shown in Eq. 2.

$$B = KERNEL * A \tag{2}$$

Here '*' operator represents convolution operator. It is to be noted that the kernel formed will be symmetric.

4.2 Intensity Gradient

Different operators are generated to give the first order derivatives in the horizontal and vertical directions as (Gx) and (Gy) correspondingly. Form these derivatives the edge gradient is calculated using the hypot function as shown in function 3,

$$G = \sqrt{G_x^2 + G_y^2} \qquad (3)$$

The direction of the edge is calculated using the atan2 function which is shown in Eq. 4,

$$\theta = atan2\left(G_y, G_x\right) \qquad (4)$$

The edge direction angle may be anything within the range of $0°$ to $180°$. But it is rounded to one of the four direction representing angles.

4.3 Non-maximum Suppression

This a famous technique followed for edge thinning. This technique suppresses all the gradient values except the local maximum to zero, and thus given the name "Non-maximum Suppression". This leaves back only the spots with sharpest variation of intensity value. Then for each pixel in the gradient image, the following algorithm is applied:

- The current pixel's edge strength is compared with the edge strength of the pixels in both positive and negative gradient directions.
- If the current pixel edge strength is largest when comparing with the other pixels in the same direction present in the mask, then the value will be preserved, otherwise the value will be suppressed.

4.4 Double Threshold

The unnecessary pixels that may bother the image in any other form must be removed. To remove these spurious pixels filter out the edge pixels that have weak gradient value and retain the pixels with higher gradient value. Thus to clarify the edge pixels two threshold values are set. The two values are high threshold value and low threshold value. The pixel values less than the low threshold is suppressed. This two threshold values are determined empirically and are should be defined for every image while processing.

4.5 Hysterisis

This is a technique for tracking the edges. Whereas the weak edge pixels may be extracted from the edge or any other variation in noise or color. So for further accuracy of result the weak edge pixels that are not extracted from the edge must be removed. So to find whether the weak edge pixel is extracted from the edge or from a noise we perform hysteresis. It uses the criteria that all the pixels extracted from an edge will be connected while the pixels out of noise will be unconnected, irrespective of weak or strong pixel [2, 8].

5 Sobel

The Sobel edge detector is mostly known as Sobel operator or Sobel filter. Irwin Sobel presented this operator, as his idea of an "Isotropic 3 × 3 Image Gradient Operator". It is a discrete differential operator. It computes and approximates the gradient of the image intensity function. Now at every point the result will be its corresponding gradient vector or the vector's normal. This operator has a small and separable filter with an integer value. The main step of this operator is convoluting this filter is in the horizontal and vertical direction [3, 9].

5.1 Gradient Approximation

The operator uses two 3 × 3 kernels to calculate the derivative approximations. One kernel is used for horizontal changes, while the other is for vertical changes. We convolute the kernels separately and get two images Gx and Gy. Gx is an image which at each point contains the approximated horizontal derivative. Similarly Gy is an image with vertical derivative approximation at each point. These images are computed as shown in Eq. 1A.

$$G_y = \begin{bmatrix} -1 & -2 & -1 \\ 0 & 0 & 0 \\ +1 & +2 & +1 \end{bmatrix} * A \text{ and } G_x = \begin{bmatrix} -1 & 0 & +1 \\ -2 & 0 & +2 \\ -1 & 0 & +1 \end{bmatrix} * A \tag{1A}$$

Here $*$ notes the two-dimensional convolution operator.

The kernel can be decomposed into averaging and differential kernels. So Gx can be written as shown in Eq. 5.

$$\begin{bmatrix} -1 & 0 & +1 \\ -2 & 0 & +2 \\ -1 & 0 & +1 \end{bmatrix} = \begin{bmatrix} 1 \\ 2 \\ 1 \end{bmatrix} [-1 \quad 0 \quad +1] \tag{5}$$

Similarly Gy can be decomposed into two kernels as expressed in 6.

$$
\begin{bmatrix} -1 & -2 & -1 \\ 0 & 0 & 0 \\ +1 & +2 & +1 \end{bmatrix} = \begin{bmatrix} -1 \\ 0 \\ +1 \end{bmatrix} \begin{bmatrix} 1 & 2 & 1 \end{bmatrix} \tag{6}
$$

Here the x-coordinate is defined to be increasing in "right" direction and the y-coordinate increases in "down" direction. The gradient approximation at the points of the image is combined to form the resulting gradient magnitude and direction. The gradient magnitude is found using formula from Eq. 3. The direction is calculated using the formula from Eq. 4. Since the kernels are separable the gradient calculation can be further simplified as shown in Eq. 7.

$$
G_x = \begin{bmatrix} 1 \\ 2 \\ 1 \end{bmatrix} * (\begin{bmatrix} 1 & 0 & -1 \end{bmatrix} * A) \text{ and } G_y = \begin{bmatrix} 1 \\ 0 \\ -1 \end{bmatrix} * (\begin{bmatrix} 1 & 2 & 1 \end{bmatrix} * A) \tag{7}
$$

The Sobel operator has two independent steps:

- Using a triangle filter to smooth the image in the direction perpendicular to the derivative direction. The triangular filter is given in Eq. 8.

$$
h(-1) = 1, h(0) = 2, h(1) = 1 \tag{8}
$$

- Along the derivative direction perform central difference. This equation set is explained in Eq. 9.

$$
h'(-1) = 1, h'(0) = 0, h'(1) = -1 \tag{9}
$$

6 Robert

Robert edge detector is more commonly referred as Robert cross. It is one of the former edge detection algorithm proposed. It was proposed by Lawrence Robert in 1963. It is a differential operator that approximates the gradient of the image through discrete differentiation. To achieve it the difference between the diagonally adjacent pixels are found and the sum of their squares is calculated.

- Robert's idea of an edge detector was:
- It should produce well-defined edges.
- The background noise must be reduced as much as possible.
- The intensity of the edge resulted must be closer to human recognition.

Along with these criteria and the psychophysical theory he proposed these set of Eqs. 11 and 12.

$$y_{i,j} = \sqrt{x_{i,j}} \tag{10}$$

$$z_{i,j} = \sqrt{(y_{i,j} - y_{i+1,j+1})^2 + (y_{i+1,j} + y_{i,j+1})^2} \tag{11}$$

Here x is pixel's initial intensity value, z is the derivative that is computed and i,j represent the pixels location within the image.

After the above operations the changes in intensity will be highlighted in a diagonal direction. The kernel used is very small and simple and contains only integers. Hence computation is very easy. A major drawback of the algorithm is that it is affected greatly by noise [6, 9].

6.1 Mathematical Execution

To perform Robert edge detection the image is convoluted with two 2×2 kernels in Eq. 13.

$$\begin{bmatrix} +1 & 0 \\ 0 & -1 \end{bmatrix} and \begin{bmatrix} 0 & +1 \\ -1 & 0 \end{bmatrix} \tag{12}$$

This forms two separate images Gx and Gy. Any point in the image I(x, y) will have corresponding points Gx(x, y) and Gy(x, y). Then the gradient at any point can be calculated using the formula in Eq. 14. This equation is also a form of Eq. 3. The gradient direction is calculated using the formula in Eq. 15. This is also a form of Eq. 4.

$$\nabla I(x, y) = G(x, y) = \sqrt{G_x(x, y)^2 + G_y(x, y)^2} \tag{13}$$

$$\theta(x, y) = arctan\left(\frac{G_y(x, y)}{G_x(x, y)}\right) \tag{14}$$

7 Prewitt

The Prewitt edge detector is mostly known as prewitt operator. This operator was developed by Judith M.S. Prewitt. It is a discrete differential operator. It computes an approximated gradient of the image intensity function. This operator results in transformation of every point in the image into its gradient vector or gradient's normal. This operator is based on convolving the image with a filter. It is to be noted that the filter is small, separable and integer valued. The convolving is done in both horizontal and vertical directions.

The main step of the operator is calculating the gradient of the image intensity at every point of the image. This gradient gives the direction of maximum possible increase in the intensity and the rate of change in that direction. The change in intensity is from light to dark. The results shows the extent of smoothness in the change at every point. Thus it shows how possibly that point is on an edge also the possible orientation of the edge [6, 9].

7.1 Mathematical Generation

The operator also uses two 3 × 3 kernels. These two are convolved with the image to calculate the approximated derivative for changes in horizontal direction and vertical direction separately. The computation is explained in Eq. 1B.

$$G_y = \begin{bmatrix} -1 & 0 & +1 \\ -1 & 0 & +1 \\ -1 & 0 & +1 \end{bmatrix} * A \text{ and } G_x = \begin{bmatrix} -1 & -1 & -1 \\ 0 & 0 & 0 \\ -1 & +1 & +1 \end{bmatrix} * A \qquad (1B)$$

Here Gx and Gy are the new derivatives. A is the source image, $*$ represents the two-dimensional convolution operator.

The Prewitt kernel can be decomposed into products of averaging and differential kernel. This will compute the gradient along with smoothing it. As it can be decomposed it is referred as separable filter.

Gx can be decomposed as detailed in Eq. 16.

$$\begin{bmatrix} -1 & 0 & +1 \\ -1 & 0 & +1 \\ -1 & 0 & +1 \end{bmatrix} = \begin{bmatrix} 1 \\ 1 \\ 1 \end{bmatrix} \begin{bmatrix} -1 & 0 & 1 \end{bmatrix} \qquad (15)$$

while Gy can be decomposed as explained in Eq. 17,

$$\begin{bmatrix} -1 & -1 & -1 \\ 0 & 0 & 0 \\ +1 & +1 & +1 \end{bmatrix} = \begin{bmatrix} -1 \\ 0 \\ 1 \end{bmatrix} \begin{bmatrix} 1 & 1 & 1 \end{bmatrix} \qquad (16)$$

Here the x-coordinate is defined as increasing right while the y-coordinate is defined as increasing down. Now the resulting gradient approximation is calculated using the same formula from Eq. 3. Gradient's direction is given by the same Eq. 4. Here θ is 0° for vertical edge, darker on right, while 90° for horizontal edge, darker down.

8 Proposed Idea

It has been stated that based on the intensive literature survey carried out, it is strongly believed that there is a need for a hybrid algorithm. In many research works conducted it had been stated that Canny gives the best result. But there may be a possibility of change in result accuracy. The idea is to enhance the outputs of various available algorithms [5].

8.1 Hybrid Algorithms

Based on the above cited results we use canny edge detector to enhance the output of the other algorithms [1, 7]. Thus we form:

- Hybrid Sobel
- Hybrid Robert
- Hybrid Prewitt

The initial step for creating the hybrid operators is to create the canny operator. So using the standard inputs of canny operator in the d2dgauss function the standard gauss filter kernels are generated.

The standard parameters for x-direction kernel are N1 = 10, Sigma1 = 1, N2 = 10, Sigma2 = 1, Theta = pi/2. Figure 1 shows the resulting kernel.

$$
g_x =
\begin{bmatrix}
0.0000 & 0.0000 & 0.0000 & 0.0000 & 0.0000 & -0.0000 & -0.0000 & -0.0000 & -0.0000 & -0.0000 \\
0.0000 & 0.0000 & 0.0002 & 0.0008 & 0.0008 & -0.0008 & -0.0008 & -0.0002 & -0.0000 & -0.0000 \\
0.0000 & 0.0003 & 0.0038 & 0.0171 & 0.0155 & -0.0155 & -0.0171 & -0.0038 & -0.0003 & -0.0000 \\
0.0000 & 0.0020 & 0.0284 & 0.1260 & 0.1142 & -0.1142 & -0.1260 & -0.0284 & -0.0020 & -0.0000 \\
0.0001 & 0.0054 & 0.0773 & 0.3426 & 0.3104 & -0.3104 & -0.3426 & -0.0773 & -0.0054 & -0.0001 \\
0.0001 & 0.0054 & 0.0773 & 0.3426 & 0.3104 & -0.3104 & -0.3426 & -0.0773 & -0.0054 & -0.0001 \\
0.0000 & 0.0020 & 0.0284 & 0.1260 & 0.1142 & -0.1142 & -0.1260 & -0.0284 & -0.0020 & -0.0000 \\
0.0000 & 0.0003 & 0.0038 & 0.0171 & 0.0155 & -0.0155 & -0.0171 & -0.0038 & -0.0003 & -0.0000 \\
0.0000 & 0.0000 & 0.0002 & 0.0008 & 0.0008 & -0.0008 & -0.0008 & -0.0002 & -0.0000 & -0.0000 \\
0.0000 & 0.0000 & 0.0000 & 0.0000 & 0.0000 & -0.0000 & -0.0000 & -0.0000 & -0.0000 & -0.0000
\end{bmatrix}
$$

Fig. 1. x-direction kernel

Similarly the parameters for y-direction were given as N1 = 10, Sigma1 = 1, N2 = 10, Sigma2 = 1, Theta = 0 and the kernel showed in Fig. 2 was obtained.

$$
g_y =
\begin{bmatrix}
0.0000 & 0.0000 & 0.0000 & 0.0000 & 0.0001 & 0.0001 & 0.0000 & 0.0000 & 0.0000 & 0.0000 \\
0.0000 & 0.0000 & 0.0003 & 0.0020 & 0.0054 & 0.0054 & 0.0020 & 0.0003 & 0.0000 & 0.0000 \\
0.0000 & 0.0002 & 0.0038 & 0.0284 & 0.0773 & 0.0773 & 0.0284 & 0.0038 & 0.0002 & 0.0000 \\
0.0000 & 0.0008 & 0.0171 & 0.1260 & 0.3426 & 0.3426 & 0.1260 & 0.0171 & 0.0008 & 0.0000 \\
0.0000 & 0.0008 & 0.0155 & 0.1142 & 0.3104 & 0.3104 & 0.1142 & 0.0155 & 0.0008 & 0.0000 \\
-0.0000 & -0.0008 & -0.0155 & -0.1142 & -0.3104 & -0.3104 & -0.1142 & -0.0155 & -0.0008 & -0.0000 \\
-0.0000 & -0.0008 & -0.0171 & -0.1260 & -0.3426 & -0.3426 & -0.1260 & -0.0171 & -0.0008 & -0.0000 \\
-0.0000 & -0.0002 & -0.0038 & -0.0284 & -0.0773 & -0.0773 & -0.0284 & -0.0038 & -0.0002 & -0.0000 \\
-0.0000 & -0.0000 & -0.0003 & -0.0020 & -0.0054 & -0.0054 & -0.0020 & -0.0003 & -0.0000 & -0.0000 \\
-0.0000 & -0.0000 & -0.0000 & -0.0000 & -0.0001 & -0.0001 & -0.0000 & -0.0000 & -0.0000 & -0.0000
\end{bmatrix}
$$

Fig. 2. y-direction kernel

Using this kernel the computation is done further as follows.

8.1.1 Hybrid Sobel

The basic Sobel algorithm is able to detect the edges but not as efficiently as Canny. So the sobel operator is modified by convoluting the result with canny operator as shown in derivation 2A.

$$
G_x = \begin{bmatrix} 1 \\ 2 \\ 1 \end{bmatrix} * \left(\begin{bmatrix} 1 & 0 & -1 \end{bmatrix} * A \right) \text{ and } G_y = \begin{bmatrix} 1 \\ 0 \\ -1 \end{bmatrix} * \left(\begin{bmatrix} 1 & 2 & 1 \end{bmatrix} * A \right) \text{ From Eq. 7}
$$

$$G = \sqrt{G_x^2 + G_y^2} \text{ From Eq. 3}$$

$$G = \sqrt{\left(\begin{bmatrix} 1 \\ 2 \\ 1 \end{bmatrix} * ([1 \quad 0 \quad -1] * A)\right)^2 + \left(\begin{bmatrix} 1 \\ 0 \\ -1 \end{bmatrix} * ([1 \quad 2 \quad 1] * A)\right)^2}$$

Combining Eqs. (7) and (3)

Now the new operator formed by convolving again is,

$$G_{1x} = G_x * \left(\sqrt{\left(\begin{bmatrix} 1 \\ 2 \\ 1 \end{bmatrix} * ([1 \quad 0 \quad -1] * A)\right)^2 + \left(\begin{bmatrix} 1 \\ 0 \\ -1 \end{bmatrix} * ([1 \quad 2 \quad 1] * A)\right)^2} \right)$$

(2A1)

$$G_{1y} = G_y * \left(\sqrt{\left(\begin{bmatrix} 1 \\ 2 \\ 1 \end{bmatrix} * ([1 \quad 0 \quad -1] * A)\right)^2 + \left(\begin{bmatrix} 1 \\ 0 \\ -1 \end{bmatrix} * ([1 \quad 2 \quad 1] * A)\right)^2} \right)$$

(2A2)

$$G_1 = \sqrt{G_{1x}^2 + G_{1y}^2} \text{ Altering Eq. 3}$$

$$\theta = atan2(G_{1y}, G_{1x}) \text{ Altering Eq. 4}$$

Here also A is the input image.

It is to be noted that once G1x and G1y are found the magnitude of the gradient vector of the image is found by same hypot function. Also the gradient direction is found using the arctan function. This operator is found to give better results compared to canny's output in many cases.

8.1.2 Hybrid Robert

Here also same mathematical methods were used. Robert uses a special small 2×2 kernel. Then the resultant kernels will be as shown in equation set 2B.

$$G_{1x} = G_x * \left(\sqrt{\left(\begin{bmatrix} +1 & 0 \\ 0 & -1 \end{bmatrix} * A\right)^2 + \left(\begin{bmatrix} 0 & +1 \\ -1 & 0 \end{bmatrix} * A\right)^2} \right) \qquad (2B1)$$

$$G_{1y} = G_y * \left(\sqrt{\left(\begin{bmatrix} +1 & 0 \\ 0 & -1 \end{bmatrix} * A\right)^2 + \left(\begin{bmatrix} 0 & +1 \\ -1 & 0 \end{bmatrix} * A\right)^2} \right) \qquad (2B2)$$

Here also A represents the input image.

After this the same procedure of calculating the gradient magnitude and direction is done using the same modified Eqs. 18 and 19 in derivation 2A. This modification showed good improvement in Roberts output which has been proven in the dataset analysis later.

8.1.3   Hybrid Prewitt

Similar to hybrid Sobel the 3×3 kernel of original Prewitt operator is convolved with the Canny's kernel. This is also based on same mathematical properties and proofs. The resulting G1x and G1y are expressed in the equation set 2C.

$$G_{1x} = G_x * \left(\sqrt{ \left(\begin{bmatrix} 1 \\ 1 \\ 1 \end{bmatrix} * ([-1 \quad 0 \quad 1] * A) \right)^2 + \left(\begin{bmatrix} -1 \\ 0 \\ 1 \end{bmatrix} * ([1 \quad 1 \quad 1] * A) \right)^2 } \right)$$

(2C1)

$$G_{1y} = G_y * \left(\sqrt{ \left(\begin{bmatrix} 1 \\ 1 \\ 1 \end{bmatrix} * ([-1 \quad 0 \quad 1] * A) \right)^2 + \left(\begin{bmatrix} -1 \\ 0 \\ 1 \end{bmatrix} * ([1 \quad 1 \quad 1] * A) \right)^2 } \right)$$

(2C2)

Here also A refers the input image.

Similarly after G1x and G1y are formed the gradient vector magnitude and its direction are calculated using the same modified Eqs. 3 and 4 in derivation 2A. This operator was also found to be better than Canny in several cases.

9   Proposed Algorithm

This algorithm is based on the original algorithm with the hybrid operators. This algorithm has been implemented in matlab. Figure 3 shows the flow chart of the algorithm proposed.

*Absolute difference1 = mod (average resemblance measure of original algorithms – resemblance measure of Canny edge detector's output).

Absolute difference2 = mod (average resemblance measure of hybrid algorithms – resemblance measure of Canny edge detector's output).

Absolute difference3 = mod (average resemblance measure of original algorithms – respective resemblance measures of the outputs)

Absolute difference3 = mod (average resemblance measure of hybrid algorithms – respective resemblance measures of the outputs)

The algorithm is briefed as:

- Filtering
- Applying the operator

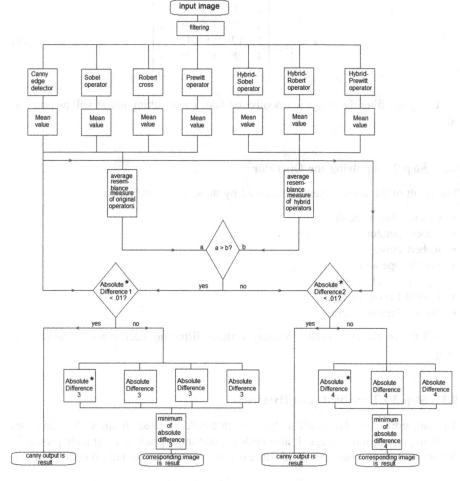

Fig. 3. Flow chart of proposed algorithm

- Threshold and hysteresis
- Enhancing the results
- Finding the mean values
- Computing the output

9.1 Step 1 - Filtering

Based on Canny's suggestion a standard 5×5 Gaussian filter is formed using the same formula in Eq. 1.

The symmetric filter formed is shown in expression 18,

$$H = \frac{1}{159} \begin{bmatrix} 2 & 4 & 5 & 4 & 2 \\ 4 & 9 & 12 & 9 & 4 \\ 5 & 12 & 15 & 12 & 5 \\ 4 & 9 & 12 & 9 & 4 \\ 2 & 4 & 5 & 4 & 2 \end{bmatrix} \qquad (18)$$

Using this filter the image is smoothened first. The resulting image will be prone to edge noise.

9.2 Step 2 - Applying the Operator

The result of the above filter is processed by these seven operators:

- Canny edge detector
- Sobel operator
- Robert cross
- Prewitt operator
- Hybrid-Sobel operator
- Hybrid-Robert operator
- Hybrid-Prewitt operator

All the results are stored separately without disturbing each other or the original image.

9.3 Step 3 - Thresholds and Hysteresis

To standardize the thresholds a dynamic threshold is used. It finds the maximum gradient present in the image. It also finds the minimum gradient magnitude present in the image. Now a single threshold value is formed using the formula shown in Eq. 19,

$$\text{Threshold} = (\text{alpha} * (\text{max} - \text{min})) + \text{min} \qquad (19)$$

Here a standard value for alpha is to be used, alpha = 0.1.

After the threshold is calculated and applied to the image, Hysteresis is done. The process of hysteresis is the same as discussed above. It is to be noted that the step should be done for all seven outputs separately.

9.4 Step 4 - Enhancing

The above outputs are sharpened using the edge enhancement technique. As discussed earlier this will increase the clarity of the image. A famous method, unsharp masking is done for enhancing the output. Now the improvements in the results may be observed directly.

9.5   Step 5 - Finding Mean Values

Once all seven outputs are processed and ready, the next work is to calculate the mean values of all the outputs and the input image. Now all the images are represented as a single floating point number. Each image is associated with the corresponding mean value.

9.6   Step 6 - Computing the Output

Once the mean values are ready the amount of resemblance is checked. As the output has only edges the amount of resemblance will help in finding the amount of edge present in each output. But the noise present in the output may also affect the amount of resemblance. From the resemblance amount the result will be computed and given. The resemblance and result are computed as follows:

- The mean values of all the outputs are subtracted from the mean value of the original image.
- The values present is the amount of data missing. This a measure of the amount of resemblance. The least value represents the largest amount of data present (resembling). It is to be noted that the value is directly proportional to the amount of resemblance.
- The average of all the resemblance measures of the original algorithms and the Hybrid operators are calculated separately.
- The minimum of the two averages are selected. The minimum average is selected based on the same fact that the values are proportional to the amount of resemblance. Thus the most matching pair is selected. In most of the cases this will be the hybrid operator's average.
- Now the absolute difference between the selected average and performance measure of canny edge detector's output is calculated.
- If the value is less than 0.01, then canny's output is given as result. This is decision is based on the fact that Canny is more efficient than other algorithms and the value 0.01 is negligible [5].
- If the value is greater than 0.01, then from the corresponding resemblance values (four for original algorithm average and three for hybrid algorithm average), the value closest to the selected average is taken out. Similarly here the value closest will be the most resembling. The highest of the group is not always the closest because of the effect of noise. The image associated to the value taken out is the required result.

10   Dataset Analysis

Based on the literature survey the original algorithms are analyzed by many researchers and stated that, Canny is the best. But all the analysis were done only a closed set. So here a range of different datasets are analyzed. The analysis were done on the original algorithm along with the hybrid algorithms. A varied range of datasets are selected for better analysis of result. The range is based on noise and minute edge structures.

10.1 Landscape

A mountain range is selected. All the seven results were compared. The result was from the Hybrid-Sobel algorithm. A survey was conducted to a group of people and the result was found to be in coincidence with most of the people's opinion. Figure 4 shows all the seven outputs of the mountain range from the operators.

Fig. 4. Mountain range edge outputs from all seven operators

10.2 Insects and Small Objects

A butterfly with more minute patterns are selected. Result was the output of Hybrid-Prewitt algorithm. This was 85% in coincidence with the people's result. The result is found to be only collections of non-continuous dots. Research is further done to enhance the algorithms to get satisfactory results in even such worst conditions. Figure 5 shows all the seven outputs of the butterfly from the operators.

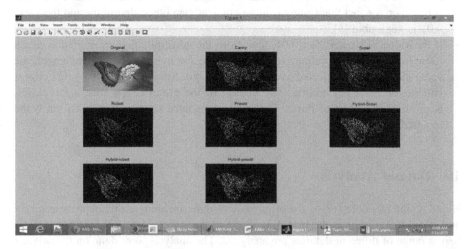

Fig. 5. Butterfly edge outputs from all seven operators

10.3 Animal

For this dataset also, to test a worst case, a tiger with a lot of stripes was selected. And the result was from canny operator. And this result was satisfactory. Figure 6 shows all the seven outputs of the tiger from the operators.

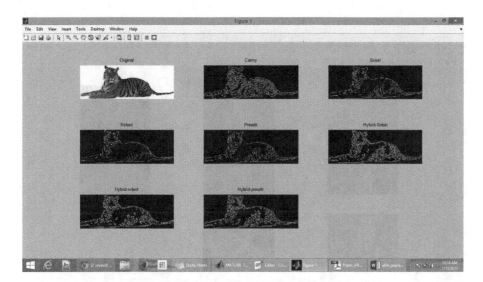

Fig. 6. Tiger edge outputs from all seven operators

10.4 Human Being

A baby's picture was selected for this datasets. The results were varied in a broad range. The result was given by Hybrid-Robert operator. Figure 7 shows all the seven outputs of the baby from the operators.

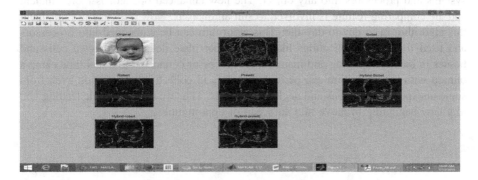

Fig. 7. Baby edge outputs from all seven operators

10.5   Buildings

A tall building with many windows and glass works were selected. This resulted in worse outline, especially in the original algorithm. The output of Hybrid-Prewitt operator was given as result. Figure 8 shows all the seven outputs of the building from the operators.

Fig. 8. Building edge outputs from all seven operators

11   Conclusion

After the extensive study done on the edge detection techniques it is important to note that all operators are simple. They can be executed easily in software and hardware like ARM/ × 86 etc., with less cost. The world of artificial intelligence has embedded the systems and processors into any object. The new enhanced operator is more efficient and the comparing feature is involved to suit all datasets. Now it is time to implement this algorithm in to day-to-day life. This algorithm can be embedded into processors and used for many applications like, car number plate detection, affected cells and tissues in medical imaging and many more. Processor connected to a real time camera burned with this algorithm can perform wonders in collecting information even from some unexpected situations and negligible objects. Thus we end the article leading you to a new beginning to move into an ocean of opportunities.

References

1. Vasudevan, K., et al.: Automotive image processing technique using cannys edge detector. IJEST 2(7), 2632–2644 (2010)
2. Ding, L., Goshtasby, A.: On the Canny edge detector. Pattern Recogn. **34**(3), 721–725 (2001)
3. James, M.: An introduction to edge detection: the sobel edge detector, pp. 969–974 (2002)
4. Maini, R., Aggarwal, H.: Study and comparison of various image edge detection techniques. Int. J. Image Process. (IJIP) **3**(1), 1–11 (2009)
5. Davis, L.S.: A survey of edge detection techniques. Comput. Graphics Image Process. **4**(3), 248–270 (1975)
6. Peli, T., Malah, D.: A study of edge detection algorithms. Comput. Graphics Image Process. **20**(1), 1–21 (1982)
7. Juneja, M., Sandhu, P.S.: Performance evaluation of edge detection techniques for images in spatial domain. Methodology **1**(5), 614–621 (2009)
8. Article Canny edge Detector. en.wikipedia.org/wiki/Canny_edge_detector. Accessed 08 July 2015
9. Bill Green, Tutorial Edge detection Tutorial (2002). http://dasl.mem.drexel.edu/alumni/bGreen/www.pages.drexel.edu/_weg22/edge.html. Accessed 09 July 2015

Malware Detection Using Higher Order Statistical Parameters

Easwaramoorthy Arul[✉] and Venugopal Manikandan

Coimbatore Institute of Technology, Coimbatore 641014, Tamilnadu, India
arulcitit@gmail.com

Abstract. Malware holds an important place in system performance degradation and information embezzling from the victim system. Most of the malware writers choose their path to reach the victim system through the internet, infected browsers, injected files, memory devices, etc., highly obscured malwares evade the automated tools installed in the victim. Once the victim system gets affected by the malware, executable processes are controlled by malware. In this paper, an algorithm has been developed to identify the malware using image processing. The malware detection process has three phases. In first phase, the files (.exe) are converted into a gray scale image. The binary values of corresponding files are converted into 8 - bit gray scale intensity value. The band pass frequency of gray scale image is computed in second phase. In the final phase, third and fourth order statistical parameter such as skewness and kurtosis are calculated at the each sub region of band pass frequency image. The region which has the highest skewness and kurtosis value is marked as the malware file. The detection performance of the proposed method has been evaluated by using 1300 portable executable files. The detection method has a true positive ratio of 93.33% with 0.1 false positives. Preliminary results indicate that the proposed algorithm is better than other conventional malware detection methods.

Keywords: Malware · Portable executable files (.exe) · Band pass frequency · Skewness · Kurtosis

1 Introduction

Malwares are the malicious programs which are created by the hackers to steal the information from the victim system. Few malwares can be easily detected by the automated tools but these tools fail to identify obscured malwares [1]. They enter into the system through the internet, infected browsers, injected files, memory devices, etc. These malwares degrade the system performance, utilizing the network bandwidth and stealing the information. Windows os X version is a widely used operating system and many malware writers are developing malicious files which are mostly in the format of portable executable (.exe). These files are spread over the internet, infected browsers, injected files, memory devices, etc., to hack the victim.

S. Subramanian et al. (Eds.): CSI 2016, CCIS 679, pp. 42–56, 2016.
DOI: 10.1007/978-981-10-3274-5_4

The portable executable programs run in the windows os X version [Fig. 1 General Architecture of windows] with valid kernel and user system calls [2,3]. The possible ways through which a malware enters into the victim system are classified as follows: Injected browsers, Injected websites, Injected files download, Proxy based firewalls, Infected USB devices, remote access and sharing/folder sharing. Injected browsers, these are browsers which are compromised by the malware [4]. When the browser starts to run, it will automatically load the hidden malicious file and utilize the internet bandwidth to steal the data and send it to the hacker. The affected browser\injected browser will starts to run like xxx.exe, 308.exe and steal the master passwords from the browser. Injected websites: the websites are compromised by the malware developer by injecting a piece of malicious code (Ex: .php, .aspx sites), so when the user hits the page from browser, the malware gets an easy way to get into the host. When the page loads in the victim machines browser, it will be automatically downloaded into the system. The compromised sites are embedded by a scripting code into the page, when page gets loaded the script will download the malicious file from the remote system into the victim. Proxy based firewalls: Many of the well-known or common service providers are provide services via many proxy servers, which will in turn update/request from primary server. So, the malware writers taking advantage of the proxy servers [5] try to send their malware code to reach many systems using multicasting. The Single attacked proxy servers its multicast the malwares code and reaches the many online systems.

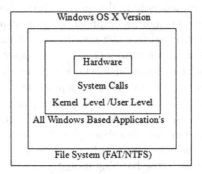

Fig. 1. Windows file system architecture

Infected USB devices: the common portable devices are the main focus of malwares writers to reach the personal computer. Once the infected USB devices connect into the system, it automatically runs the malware code and the malware immediately takes control of the system via affecting the kernel files. The Pen drives are most commonly affected devices, when connected into the system, it automatically runs the autorun.ini which itself have the shell script to run the hidden malware code. Remote access and sharing/folder sharing: Many of organizational computers are commonly connected via any one of the networks [6]. Example: LAN (Local Area Network), WAN (Widal Area Network),

Ad Hoc network, etc. When any of the systems in the network are affect with malware, then the malware gets easily to the other system via remote access or file sharing. When the personal computer enables the remote file sharing, the online hacker easily downloads the malware code into the shared folder.

The paper is divided into four different sections, Sect. 2 describes literature survey on the existing malware detection techniques, Sect. 3 describes the proposed work based on higher order statistical parameter, Sect. 4 describes about the results and comparison of the proposed method with various existing methods and Sect. 5 concludes the work and discuss its future scope.

2   Literature Survey

The challenges being thrown to the modern world by malicious softwares (malware) and need to counteract them are becoming increasingly imminent. This is true in spite of the great improvements in the efficacy of procedures of malware propagation detection, analysis and updating, the bases of signatures and detection rules. The focus of this problem is to look for more reliable heuristic detection methods. These methods aim at recognizing of new malicious programs which cannot be detected by using traditional signature- and rule-based detection techniques which are oriented to search for concrete malware samples and families. These heuristic methods provide immunity and defense against targeted and zero-day attacks, since the rate of detecting such relatively new types of threats by traditional techniques is not satisfactory. Data mining methods are used for constructing heuristic malware detectors. The approach described below differs from others through its emphasis on processing static, positionally dependent features which considers the specificities of the objects file format, which is potentially a malware container [7]. The paper explains the investigation and realization of the common methodology for design of Data Mining-based malware detectors using positionally dependent static information. The given approach does not guarantee absolute accuracy of malware detection. But it can be effective to make decision by further processing the object and to detect particular families of executable. For example, this approach can be used in a task of automating the identification of obfuscation.

A major security threat to the banking industry worldwide is cyber fraud and malware is one of the manifestations of cyber frauds. Malware propagators make use of Application Programming Interface (API) calls to perpetrate these crimes. Malware detection by text and data mining provides a static analysis method to detect Malware based on API call sequences using text and data mining in tandem [8]. In this paper the dataset available at CSMINING group is analyzed, text mining is employed to extract features from the dataset consisting of a series of API calls. Also, mutual information is invoked for feature selection, and then resorted to over-sampling to balance the data set. Finally, various data mining techniques such as Decision Tree (DT), Multi-Layer Perceptron (MLP), Support Vector Machine (SVM), Probabilistic Neural Network (PNN) and Group Method for Data Handling (GMDH) used to identify malware. Throughout the paper, 10-fold cross validation technique is used for testing the techniques. It is noteworthy to mention that SVM and OCSVM achieved 100% sensitivity after balancing the dataset.

As the internet has evolved, the number of malicious software, or malware, distributed especially for monetary profits, is exponentially increasing, and malware authors are developing malware variants even using various automated tools and techniques. Automated tools and methods reuse some modules to develop malware variants, so these reused modules can be used to classify malware or to identify malware families. Therefore, similarities that exist among malware variants can be analyzed and used for malware variant detections and the family classification. Malware analysis using visualized images and entropy graphs proposes a new malware family classification method by converting binary files into images and entropy graphs [9]. To analyze the malware binary files in which packing techniques are applied, the method can be extended to instruction-level analysis with the aid of dynamic analysis [10].

A few different methods have been devised by the researchers to facilitate malware analysis and one of them is through malware visualization. Malware visualization is a field that focuses on representing malware features in a form of visual cues that could be used to convey more information about a particular malware. There has been research in malware visualization but however, there seems to be lack of focus in visualizing malware behavior. Emphasis is on analyzing visualizing malware behavior and its potential benefit for malware classification. Malware Image Analysis and Classification using Support Vector Machine research depicts that malware behavior visualization can be used as a way to identify malware variants with high accuracy [11]. The proposed method concentrates leveraging both packed and unpacked malicious executable files for further research study on static analysis of malware.

Graph-based malware detection which utilizes dynamic analysis introduces a novel malware detection algorithm based on the analysis of graphs constructed from dynamically collected instruction traces of the target executable [12]. These graphs represent Markov chains, where the vertices are the instructions and the transition probabilities that are estimated by the data contained in the trace. The graph kernels are combined to create a similarity matrix between the instruction trace graphs. The resulting graph kernel measures similarity between graphs on both local and global levels. Finally, the similarity matrix is sent to a support vector machine to perform classification. This method is particularly appealing as the classifications are not based on the raw n-gram data, but rather on data representation which is used to perform classification in graph space.

Most of the existing schemes for malware detection are signature-based and so they can effectively detect known malwares, but they cannot detect variants of known malwares or new ones. Most network servers do not expect executable code in their in-bound network traffic, such as on-line shopping malls, Picasa, Youtube, Blogger, etc. Therefore, such network applications can be protected from malware infection by monitoring their ports to see if incoming packets contain any executable contents. Classification of packet contents for malware detection proposes a content-classification scheme that identifies executable content from incoming packets [13]. The proposed scheme analyzes the packet payload in two steps. It first analyzes the packet payload to see if it contains multimedia-type data (such as avi, wmv, jpg). If not, then it classifies the payload either as text-type (such as txt, jsp, asp) or executable. The proposed scheme shows a low rate of false negatives and positives (4.69% and 2.53%, respectively), the presence of inaccuracies demand further inspection to efficiently detect the occurrence of malware.

A run-time malware detection method which is based on positive selection is a supervised methodology to detect malwares [14]. A novel classification algorithm based on the idea of positive selection is proposed, which is one of the important algorithms in Artificial Immune Systems (AIS), inspired by the biological phenomenon of positive selection of T-cells. The proposed algorithm is applied to learn and classify program behavior based on I/O Request Packets (IRP). After extensive analysis, it is found that the difference between two IRP traces of the same program is that, sometimes if an IRP is not successful. The IRP will keep trying until becoming successful. So, the real difference between the two IRP traces is that some IRPs repeat some times. This difference has little effect on the results. The system does not need to be retrained in the event of installation of new applications or programs.

Most of the researchers have analysed to detect the malware based on Data Mining, optimization technique, neural network, fuzzy logic and SVM techniques. In their methods, the malwares can be efficiently detected, but different signature of malware cannot be detected by their method. Therefore, the numbers of malware missing are quite high in the existing methods. It might be possible to improve detection accuracy by introducing the skewness and kurtosis parameter in malware image analysis. The proposed system has the potential to detect malware with high detection accuracy.

3   Malware Detection

The Fig. 2 shows the flowchart of the developed malware detection system. In this system, initially, Portable Executable (PE) files are taken from database and converted into gray scale image. The portable executable is in the format of binary by default. The 8 consecutive bits are taken from the binary file and are converted into decimal value. This decimal value is equivalent to 8-bit intensity of the image. Similarly, all the bits in the binary file are converted into gray scale image. The malware regions are identified using Higher Order Statistical Parameters (HSP) such as Skewness and Kurtosis from gray scale image, which are measures of the asymmetry and impulsiveness of the distribution. Skewness is a measure of symmetry, or more precisely, the lack of symmetry. If a distribution is symmetric, the corresponding Skewness value is close to zero. The Skewness of a random variable, denoted by γ_3 is defined in Eq. (1) [16–18]

$$\gamma_3 = \frac{E\{[x - E(x)]^3\}}{(E\{[x - E(x)]^2\})^{\frac{3}{2}}} \tag{1}$$

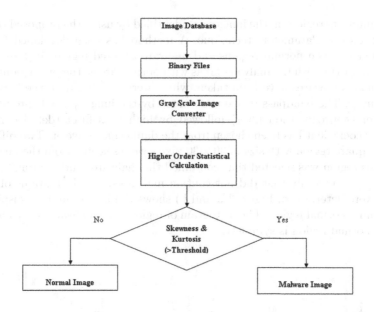

Fig. 2. Architeure of proposed higher order statistical parameters (HSP)

$E(x)$ is the mean value of x and E is the expectation operator. For a sample of N values, the sample Skewness is calculated using Eqs. (2) and (3)

$$Skewness = \frac{m_3}{m_2^{\frac{3}{2}}} \tag{2}$$

$$m_3 = \frac{\sum_{i=1}^{N}(x - \tilde{x})^3}{N} \quad and \quad m_2 = \frac{\sum_{i=1}^{N}(x - \tilde{x})^2}{N} \tag{3}$$

The Kurtosis is a measure of the heaviness of the tails in the distribution. Distribution has high Kurtosis value with sharp peaks and heavy tail. Distribution with low Kurtosis tends to have a flat top near the mean rather than a sharp peak. A Kurtosis value is low when distribution is symmetric.

The Kurtosis of a random variable, denoted by γ_4 is defined in Eq. (4) [16–18]

$$\gamma_4 = \frac{E\{[x - E(x)]^4\}}{(E\{[x - E(x)]^2\})^2} - 3 \tag{4}$$

An estimate of the Kurtosis is given in Eqs. (5) and (6)

$$Kurtosis = \frac{m_4}{m_2^2} - 3 \tag{5}$$

$$m_4 = \frac{\sum_{i=1}^{N}(x - \tilde{x})^4}{N} \quad and \quad m_2 = \frac{\sum_{i=1}^{N}(x - \tilde{x})^2}{N} \tag{6}$$

Where, $\tilde{x}$ is the mean value, m_4 is the fourth central moment, m_3 is the third central moment, and m_2 is the variance of the data.

The malware regions in the images are identified by using the proposed Higher order Statistical Parameters technique. A method has been developed for the distinction between normal regions and malware affected regions in the image. In order to distinguish the malware areas with normal areas, the two regions with and without malwares have been taken, which were selected from the bandpass of the image. The bandpass image is obtained by the image passed through low pass filter (approximation filter) followed by high pass filter (detailed filter). The filter coefficient has been chosen from the daubechies wavelet. Two different types of square region with size of 30 × 30 pixels were selected from the image. A suspicious region was selected that contained the malwares and a normal region was selected randomly that did not contain malwares. The histograms of both regions were determined. Figure 3(a and b) shows the histogram of a suspicious region and a normal region. The histogram of region with malware is asymmetric and the normal region is symmetric.

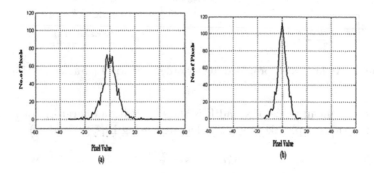

Fig. 3. (a) Histogram of regions with malware regions (b) Histogram of normal regions

The Malware region detection approach is as follows: The 30 × 30 pixels square mask is moved on image, in which Skewness value (Sk) and Kurtosis value (Ku) are computed at each single pixel displacement. Regions with high positive Skewness and Kurtosis are marked as suspicious regions. The distribution of suspicious regions is asymmetric and heavy tail, therefore the magnitude of Skewness and Kurtosis value is high in malware region.

A single row on the malware affected file image was taken, which contains the malware. Figure 4(a and c) shows the single row pixel value of malware affected region. Besides, Fig. 4(b and d) shows a single row pixel value of normal region which does not contain malware.

From the Fig. 4, it is clearly found that it is tedious task to differentiate both the regions. In order to distinguish both regions, Skewness and Kurtosis were calculated for both regions. Figure 5(a and b) shows Skewness and Kurtosis values of Figs. 4(a and b) and 5(c and d) shows Skewness and Kurtosis values of Fig. 4(c and d). From the Fig. 5, it is identified that Skewness and Kurtosis values are higher than normal regions. Third order moment (Skewness) and fourth order moment (Kurtosis) are calculated using Eqs. (2) and (5). A statistical analysis

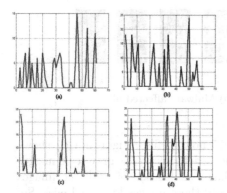 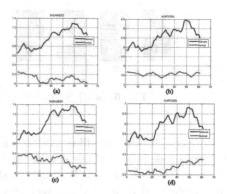

Fig. 4. Single row profile of normal and malware PE image

Fig. 5. Skewness and kurtosis values of normal and malware image files

for calculation of Skewness and Kurtosis in normal and Malware affected regions was carried out. Higher order analysis was performed on 18 random regions, of which 9 regions have malwares and 9 regions are normal regions. The Skewness and Kurtosis of various suspecting regions and normal regions are calculated and are displayed in Table 1.

Table 1. Skewness and kurtosis for suspicious regions and normal regions

Region	Skewness ($|S_K|$)	Kurtosis ($|K_u|$)	Region	Skewness ($|S_K|$)	Kurtosis ($|K_u|$)
Suspicious region 1	0.4684	4.979	Normal region −1	0.362	3.609
Suspicious region 2	0.5213	4.764	Normal region −2	0.224	3.821
Suspicious region 3	0.7971	4.939	Normal region −3	0.368	3.724
Suspicious region 4	0.4633	4.217	Normal region −4	0.142	1.682
Suspicious region 5	0.6702	4.766	Normal region −5	0.405	2.797
Suspicious region 6	0.4354	3.976	Normal region −6	0.380	3.913
Suspicious region 7	0.4326	4.605	Normal region −7	0.119	3.841
Suspicious region 8	0.5732	4.437	Normal region −8	0.391	4.518
Suspicious region 9	0.8841	4.826	Normal region −9	0.295	3.482

The regions in the malware having $|S_K| > 0.4$ and $|K_u| > 4$ are identified as suspicious regions. The above threshold values were estimated from analyzing more suspicious and normal regions. Therefore, it is concluded that higher order statistical parameters such as skewness and kurtosis are used to identify the malware affected regions.

4   Results and Comparison

The binary files which contain malware are identified by band pass frequency based Higher Order Statistical Parameter methods. Malware analysis was performed for 1300 files that were converted into gray scale images. The images

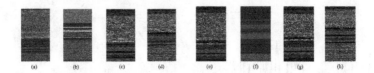

Fig. 6. (a–d) Normal files (e–h) Malware affected files

contained 345 malware files [20–22] and 955 normal files. Figure 6(a–d) shows the gray scale images of normal files and Fig. 6(d–h) shows the gray scale images of malware affected files. It is observed that visualization of both the images are similar and it is a tedious task to classify these images by naked eyes. So, the image processing systems are needed to classify these images. The proposed method can able to identify the malware affected files.

Figure 7(a–d) shows normal PE files which is not affected by any malware. Figure 7(e–h) shows malware detection output images by the proposed Higher Order Statistical method (HSP) and Fig. 7(i–l) shows detected results superimposed on normal PE Files. From the results, it is found that the normal PE files can be identified as a normal PE file accurately by the proposed method. But, the proposed algorithm also has some limitations. Figure 7(c) shows the normal PE file but the proposed method identifies this normal PE file as abnormal file. This is the limitation of the proposed method but it can be avoided by increasing the threshold value of Skewness and Kurtosis value.

Figure 8(a–d) shows PE files which is affected by malware. Figure 8(e–h) shows malware detection output images by the proposed Higher Order Statistical method and Fig. 8(i–l) shows detected results superimposed on malware affected PE files. From the experimental results, it is observed that proposed method can detect the malware in the PE Files efficiently and accurately.

The Free-Response Operating Characteristic (FROC) curve is used to evaluate the performance of malware detection method [19]. This plot provides true-positive detection ratio versus the average number of False Positives (FPs) per image. The TP ratio refers to the percentage of malware files that are truly detected and FP number/image refers to normal PE files that are wrongly identified as malware files by the proposed algorithm. True positive detection ratio refers to the malware files that are correctly detected by proposed higher order statistical method. The proposed method has been evaluated using FROC curve and it is shown in Fig. 9. The FROC curve presents a summary of the percentage of malware files truly detected and also the percentage of normal files identified correctly.

A comparison of the proposed method is performed with the malware detection methods that was proposed by Kyoung Soo Han et al. [9] and Lakshman Nataraj [15]. Kyoung Soo Han et all have developed a method for malware classification based on the entropy graph similarity [9]. In their method, binary files are converted into bitmap images by bitmap converter. After the bitmap image conversion, the entropy value of each line of bitmap is calculated and entropy

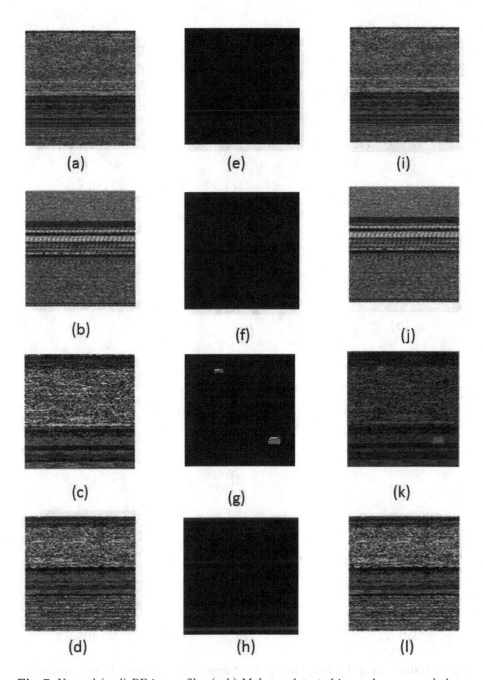

Fig. 7. Normal (a–d) PE image files (e–h) Malware detected image by proposed algorithm (i–l) Detected results superimposed on original PE files

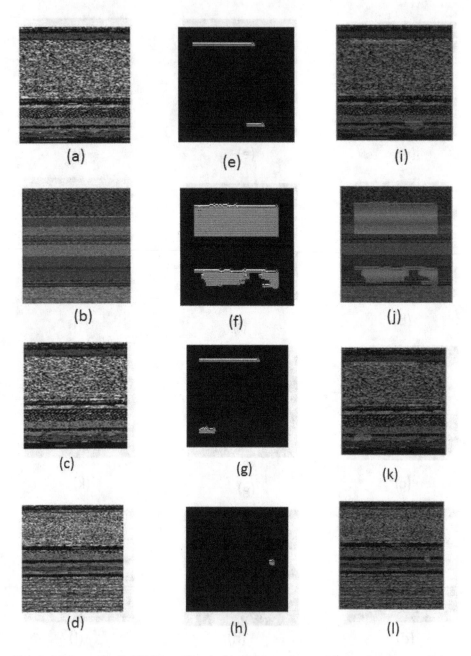

Fig. 8. Malware (a–d) PE image files (e–h) Malware detected image by proposed algorithm (i–l) Detected results superimposed on original PE files

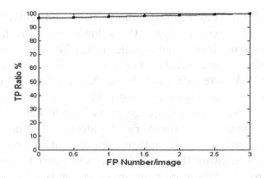

Fig. 9. FROC curve of proposed HSP for malware detection

Table 2. Comparison of true positive ratio's of various detection methods with proposed HSP

Malware family	No. files taken for analysis	No. files detected by proposed method	TP ratio (%) of proposed method	No. files detected by entropy method	TP ratio (%) of entropy method	No. files detected by texture feature method	TP ratio (%) of texture feature method
Rootkits	34	32	94.12	31	65	30	88.23
Ransomware	05	3	60	3	60	2	40
Rogue software	03	2	66.67	2	66.67	2	66.67
Spyware	16	15	93.75	12	0.75	10	62.5
Backdoors	42	38	90.48	36	85.71	32	76.19
Keyloggers	12	9	75.00	7	58.33	8	66.66
Browser Hijacker	23	22	95.65	21	91.30	20	86.95
Trojan horses	97	91	93.81	86	88.65	87	89.69
Adware	21	21	100.00	21	100.00	18	85.71
Bots	14	13	92.86	13	92.85	12	85.71
Worms	15	15	100.00	14	93.33	11	73.33
Viruses	63	61	96.83	57	93.44	55	87.30

Table 3. Comparison of true and false positive ratio's of various detection methods with proposed HSP

Methods	Number of malware images are detected	TP ratio (%)	FP detected	FP ratio (%)
Kyoung	303	87.83	135	0.14
Lakshman	287	83.18	172	0.18
Proposed HSP	322	93.33	96	0.10

Total Number of Malware Image File Taken For Analysis 345
*Total Number of Normal Image File Taken For Analysis 955

graphs based on these values are generated. These entropy graphs are stored as features and the features are used to identify malware by calculating similarities of entropy graphs. For similarity calculation, the threshold value is used and it is fixed as 0.75. They achieved about 3% false-negative rate and 3.5% false-positive rate. They have analyzed the following malware families: Worms, Backdoors, Trojans and Viruses that are files in the Windows operating system. But still false positive is high and their method cant identify some malware family (Ex: Trojan-PSW.Win32.EPS). Lakshman Nataraj has developed a method for detecting malware using image processing [15]. The binary malware is converted into 8 bit grayscale image. Image provides relevant information about the structure of the malware. Malware can be identified by image features and feature extraction based on image texture. Demerits of this method are that it can be able to detect malware that already exists in the database, but it fails to detect new malwares. Besides more malware detection is achieved by extracting more texture features from image. But both algorithms failed to detect different malware families. These two methods were compared by conducting experimental analysis for 1300 PE image files that contained 345 malware image files, of which 322 malwares were detected by the proposed method, 303 malware were detected by Kyoung entropy method and 287 malware were detected by Lakshman method. The number of false positives obtained by proposed method, Kyoung entropy, and Lakshman method were 96, 135, and172, respectively. Therefore, the proposed method has the TP ratio of 93.33% (322/345) with 0.1 (96/955) Fps per image, Kyoung entropy method has TP ratio of 87.83% (303/345) with 0.14 (135/955) Fps per image, and Lakshman method has TP ratio of 83.18% (287/345) with 0.18 (172/955) Fps per image. Tables 2 and 3 summarize the comparison of the proposed method with two existing methods. Most of the malwares such as Rootkits, Spyware, Backdoors, Browser Hijacker, Trojan horses, Bots, Viruses are correctly identified and TP ratio of this malware detection is high. Some of the malwares Ransomware, Rogue security software, Keyloggers are missed, because the texture pattern of malware affected file is similar to normal file. From the results, it is found that the proposed method is able to detect different families of malware than other methods.

5   Conclusion and Future Work

Malicious process is a major threaten to software for both system user and antivirus/antimalware products. In this paper, the malicious software in the form of portable executable (as binary was focused) are converted into gray scale image. An algorithm has been developed to assist malware detector for accurate malware detection in Portable Executable files. Malware files were identified by using band-pass frequency based method. The 30×30 pixels square mask was moved on bandpass subimage, in which Skewness and Kurtosis were computed. The sub image which had higher Skewness and Kurtosis value were marked as malware affected area and the threshold values have been chosen by analyzing the more normal and malware files. The proposed method achieved TP ratio of 93.33% with 0.1 False Positives per Image (FPI). The strength of the proposed approach is that it detects various types of malware accurately. Now a days, large volumes of files are scanned by the malware detection algorithm and it is time consuming. In future, it is planned to reduce the time delay for analyzing a single file without affecting the detection accuracy. In future, texture properties of the malware will be considered to increase the detection accuracy. The proposed algorithm elevates the signature of malicious file, which can be utilized by antivirus/antimalware products.

References

1. Chandola, V., Banerjee, A., Kumar, V.: Anomaly detection: a survey. ACM Comput. Surv. (CSUR) **41**(3), 15 (2009)
2. Pietrek, M.: An In-Depth Look into the Win32 Portable Executable, File Format, MSDN Magazine (2002). https://msdn.microsoft.com/en-IN/library/ms809762.aspx, http://www.csn.ul.ie/caolan/pub/winresdump/winresdump/doc/pefile2.html
3. https://msdn.microsoft.com/en-us/library/t0wd4t32.aspx, https://www.cs.columbia.edu/jae/4118/L02-intro2-osc-ch2.pdf
4. Hsu, F.H., Tso, C.K., Yeh, Y.C., Wang, W.J., Chen, L.H.: Browser guard: a behavior-based solution to drive-by-download attacks. IEEE J. Sel. Areas Commun. **29**(7) (2011). doi:10.1109/JSAC.2011.110811
5. Watson, M.R., Shirazi, N.U.H., Marnerides, A.K., Mauthe, A., Hutchison, D.: Malware detection in cloud computing infrastructures. IEEE Trans. Dependable Secur. Comput. doi:10.1109/TDSC.2015.2457918
6. Butler, P., Rhodes, A., Hasan, R.: MANTICORE: masking all network traffic via IP concealment with OpenVPN relaying to EC2. In: 2012 IEEE 5th International Conference on Cloud Computing (CLOUD). Print ISBN: 978-1-4673-2892-0, doi:10.1109/CLOUD.2012.29
7. Komashinskiy, D., Kotenko, I.: Malware detection by data mining techniques based on positionally dependent features. In: 2010 18th Euromicro International Conference on Parallel, Distributed, Network-Based Processing (PDP), 17–19 February 2010. ISSN: 1066-6192, doi:10.1109/PDP.2010.30
8. Sundarkumar, G.G., Ravi, V.: Malware detection by text, data mining. In: 2013 IEEE International Conference on Computational Intelligence, Computing Research (ICCIC), 26–28 December 2013. ISBN: 978-1-4799-1594-1, doi:10.1109/ICCIC.2013.6724229
9. Han, K.S., Lim, J.H., Kang, B., Im, E.G.: Malware analysis using visualized images and entropy graphs. Int. J. Inf. Secur. http://link.springer.com/article/10.1007/s10207-014-0242-0
10. Dai, J., Guha, R., Lee, J.: Efficient virus detection using dynamic instruction sequences. J. Comput. **4**(5), 405–414 (2009)
11. Makandar, A., Patrot, A.: Malware image analysis and classification using support vector machine. Int. J. Adv. Trends Comput. Sci. Eng. **4**(5), 01–03 (2015)
12. Anderson, B., Quist, D., Neil, J., Storlie, C., Lane, T.: Graph-based malware detection using dynamic analysis. J. Comput. Virol. **7**(4), 247–258 (2011). First online: 08 June 2011
13. Ahmed, I., Lhee, K.-S.: Classification of packet contents for malware detection. J. Comput. Virol. **7**, 279 (2011). First online: 22 October 2011
14. Fuyong, Z., Deyu, Q.: Run-time malware detection based on positive selection. J. Comput. Virol. **7**, 267 (2011). First online: 28 July 2011

15. Nataraj, L., Karthikeyan, S., Jacob, G., Manjunath, B.S.: Malware images: visualization and automatic classification. In: Proceedings of the 8th International Symposium on Visualization for Cyber Security, Article No. 4. ISBN: 978-1-4503-0679-9, doi:10.1145/2016904.2016908

16. Tsatsanis, M.K., Giannakis, G.B.: Object and texture classification using higher-order statistics. IEEE Trans. Pattern Anal. Mach. Intell. **14**(7), 733–750 (1992)

17. Gurcan, M.N., Yardimci, Y., Cetin, A.E., Ansari, R.: Detection of microcalcifications in mammograms using higher order statistics. IEEE Signal Process. Lett. **4**(8), 213–216 (1997)

18. Balakumaran, T., Vennila, I.: A computer aided diagnosis system for microcalcification cluster detection in digital mammogram. Int. J. Comput. Appl. **34**(1), 39–45 (2011)

19. Oliva, A., Torralba, A.: Modeling the shape of the scene: a holistic representation of the spatial envelope. Int. J. Comput. Vis. **42**(3), 145–175 (2001)

20. Kirat, D., Nataraj, L., Vigna, G., Manjunath, B.S.: SigMal: a static signal processing based malware triage. In: Proceedings 29th Annual Computer Security Applications Conference, pp. 89–98, December 2013

21. Wu, P., Guo, Q., Song, H., Tang, X.: A guess to detect the downloader-like programs. In: Proceedings of the Ninth International Symposium Distributed Computing and Applications to Business Engineering and Science, Hong Kong, China, 10–12 August 2010

22. http://www.tekdefense.com/downloads/malware-samples/, http://contagiodump.blogspot.in/2010/11/links-and-resources-for-malware-samples.html, https://tuts4you.com/download.php?list.89

Denoising Iris Image Using a Novel Wavelet Based Threshold

K. Thangavel[(⊠)] and K. Sasirekha

Department of Computer Science, Periyar University, Salem, Tamilnadu, India
drktvelu@yahoo.com, ksasirekha7@gmail.com

Abstract. The efficiency of an iris authentication system depends on the quality of the iris image. Denoising of the iris image is indispensable to get a noise free image. In this paper, a novel method is proposed to remove Gaussian noise present in the iris image using Undecimated wavelet, a threshold based on Golden Ratio and weighted median. First, decompose the input image using Stationary Wavelet Transform (SWT) and apply the modified Visushrink to the wavelet coefficients using hard and soft thresholding. Then apply inverse SWT to get the noise free image. Different kinds of wavelet filters such as db1, db2, sym2, sym4, coif2 and coif4 for different noise levels are performed. The filter db1 is outperformed. In this research, experiments have been conducted on the iris database CASIA. The Peak Signal-to-Noise Ratio (PSNR), Signal-to-Noise Ratio (SNR), Root Mean Square Error (RMSE) and Mean Square Error (MSE) have been computed and compared.

Keywords: Iris · Golden ratio · Hard threshold · Undecimated wavelet · Visushrink · Wavelet filters · Weighted median

1 Introduction

Biometric authentication refers to verifying individuals based on their physiological and behavioural characteristics. Now-a-days biometric technologies are widely used in many applications for various purposes of personal authentication. Biometric methods provide a higher level of security and are more convenient for the user than the traditional methods of personal authentication such as passwords and tokens [1]. Among all the biometrics, iris recognition can be considered as one of the most reliable and accurate method of biometric technology. The iris is an externally visible and protected organ whose unique pattern remains stable throughout adult life [2, 3].

Iris images are contaminated with Gaussian noise during its acquisition and transmission. Denoising of the iris image is a vital task before further processing to get a reliable and accurate result. So far several techniques have been developed for reducing the noises in iris image both in spatial and wavelet domain.

In spatial domain the concept behind image denoising is convolution and moving window principle. In [4], an advance denoising and smoothing technique on the captured iris images using the modified version of the anisotropic diffusion is presented. An attempt is made to retain all the properties of the original model and to enhance the performance in the presence of noise.

© Springer Nature Singapore Pte Ltd. 2016
S. Subramanian et al. (Eds.): CSI 2016, CCIS 679, pp. 57–69, 2016.
DOI: 10.1007/978-981-10-3274-5_5

An efficient approach to remove white noise present in iris image is proposed, in which phase preserving principle is held to avoid corruption of iris texture features. Importance of phase information for iris image is shown by an experiment and the method to implement phase preserving by complex Gabor wavelets is explained [5].

With Wavelet Transform gaining popularity in the last two decades various algorithms for denoising in wavelet domain were introduced. Wavelets are mathematical functions that analyze data according to scale or resolution. Wavelet transforms have become one of the most important and powerful tools for image processing [6].

Furthermore, the wavelet provides an appropriate basis for separating noisy coefficient from the image than spatial and frequency based methods. The small coefficients are more likely due to noise and large coefficients are image features. These small coefficients can be thresholded without affecting the significant features of the image. The procedure in which small coefficients are removed while others are left is called Hard Thresholding [7]. On the other hand, soft thresholding shrinks the coefficients in the image above the threshold in absolute value.

Denoising by thresholding in the wavelet domain has been developed principally by Donoho et al. [8]. VisuShrink was introduced by Donoho. It depends on the number of pixels in the image and noise variance of the image. A threshold based on Stein's Unbiased Risk Estimator (SURE) was proposed by Donoho and Johnstone [10] and is called as SureShrink. It is a combination of the VisuShrink and the SURE threshold. This method specifies a threshold value tj for each resolution level j in the wavelet transform which is referred to as level dependent thresholding. BayesShrink was proposed by Chang, Yu and Vetterli [9]. The goal of this method is to minimize the Bayesian risk, and hence its name, BayesShrink. It uses soft thresholding and is subband-dependent, which means that thresholding is done at each band of resolution in the wavelet domain.

In this paper, a computationally inexpensive yet effective method based on VisuShrink for iris image denoising using Golden Ratio (GR) and weighted median is proposed. The proposed method uses undecimated wavelet transform (also known as SWT) for decomposition and reconstruction over the conventional DWT as it is not a time-invariant transform. The different kinds of wavelet filters such as db1, db2, sym2, sym4, coif2 and coif4 have been applied to different noise levels.

The paper is organized as follows. Section 2 presents the review of wavelet based thresholding methods for image denoising. In Sect. 3, a modified VisuShrink using GR and weighted median has been proposed. Section 4 provides experimental results of the proposed method on iris images and compared with the results of the existing methods. Finally, this paper concludes with some perspectives in Sect. 5.

2 Analysis of Image Denoising in Wavelet

2.1 Undecimated Wavelet Transform

The SWT is studied over the conventional DWT as it is not a time-invariant transform [6]. During the wavelet transformation, low frequency components in the image will be filtered out as approximations and high frequency components will be filtered out as

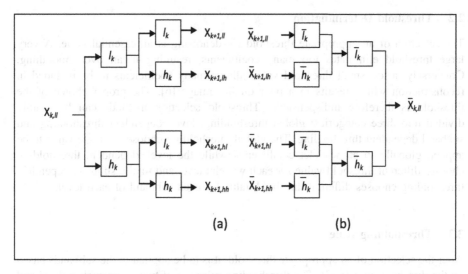

Fig. 1. One level filter bank implementation of SWT (a) Forward Pass (b) Backward Pass

details. Figure 1 depicts one level filter bank implementation of SWT. It applies high and low pass filters to the data at each level.

The SWT produces four subbands at each level of decomposition. They are approximation and detail coefficients such as horizontal, vertical and diagonal as shown in Fig. 2.

LL (Approximation)	HL (Horizontal)
LH (Vertical)	HH (Diagonal)

Fig. 2. Decomposition of an image using SWT

Here, H and L denote high and low-pass filters respectively. The LL subband is the low resolution residual consisting of low frequency components and this subband which is further split at higher levels of decomposition. After decomposition, the coefficients in the detail subbands are thresholded to remove noise and finally the denoised image is reconstructed from the thresholded subbands [5]. The steps in wavelet based image denoising are as follows:

- Decompose the noisy image using SWT.
- Threshold the wavelet coefficients using the selected threshold method.
- Reconstruct the image using ISWT to get the noise free image.

2.2　Threshold Determination

The selection of an appropriate threshold for denoising is an essential issue. A very large threshold eliminates too many coefficients, resulting in an over smoothing. Conversely, a too small threshold value allows many coefficients to be included in reconstruction which results in a poor quality image [6]. The proper choice of the threshold is therefore indispensable. Threshold selection methods can be mainly divided into three categories: global thresholding, level−dependent thresholding and subband dependent thresholding. The global thresholding chooses a single value to be applied globally to all wavelet coefficients, while the level−dependent thresholding chooses different threshold value for each wavelet level and finally subband dependent thresholding chooses different threshold value for each subband of each level.

2.3　Thresholding Rule

After the selection of an appropriate threshold, it is to be applied to the subbands based on the thresholding rule [7]. The thresholding rules are of two types such as hard and soft. These two rules are studied in the following sequel.

The hard threshold removes coefficients below a threshold value (λ) which is determined by the thresholding algorithm. This is sometimes known as "keep or kill" method since it keeps the coefficient above the threshold and kills the coefficients below the threshold value.

$$I(u,v) = u \, for \, all \, u > \lambda$$
$$= 0 \ otherwise \tag{1}$$

Where I is the image and u is the pixel value of a particular image in absolute value. Soft thresholding shrinks the coefficients in the subband above the threshold in absolute value.

$$I(u,v) = sign(u) \, \max(u - \lambda) \tag{2}$$

3　Proposed Modified Visushrink Based on Golden Ratio and Weighted Median

3.1　Visushrink

Visushrink was introduced by Donoho [8]. It is also referred to as Universal Threshold (T) and it is defined as

$$T = \sigma(\sqrt{2 * Log(N)}) \tag{3}$$

where N is the number of elements or pixels in the image and σ is the noise variance in that image, which is calculated from the diagonal subband (HH) as

$$\sigma = \frac{Median(HH)}{0.6745} \tag{4}$$

To estimate the noise level σ, Donoho used a result proposed by Frank R Hampel [11] and showed that the Median Absolute Deviation MAD (X) = |X–Median (X)| converges to 0.6745 times σ as the sample size goes to infinity. Note that the median is computed only by considering the absolute value of all pixels.

3.2 Proposed Threshold

The Visushrink is modified using Golden ratio and weighted median to improve the performance of denoising method. The Golden ratio is also called the Golden section or Golden mean [12]. The block diagram of the proposed method is shown in Fig. 3.

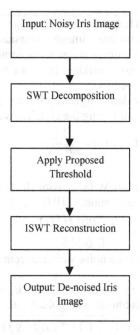

Fig. 3. Block diagram of the proposed method

In (3), instead of computing log (N) two times, the value of the golden ratio '1.618' is multiplied with log (N) in the proposed to compute the threshold (T) as follows:

$$T = \sigma(\sqrt{1.618 * Log(N)}) \tag{5}$$

Weighted median is proposed to compute the median of the high pass portion of the image instead of the conventional median given in (4). This paper adopts the following classical weight function (W) [13] for computing the weighted coefficient of the diagonal subband (HH) which is given by (6),

$$\mathbf{W}(x, y) = \frac{1}{e^{|HH(x,y)|}} \qquad (6)$$

Here x and y are the coordinates in the HH subband. The weight W will be multiplied with HH to get weighted diagonal subband as given in (7),

$$HH1(x, y) = W(x, y) * HH(x, y) \qquad (7)$$

The noise variance σ is then calculated from the weighted diagonal subband (HH1) as given in (8),

$$\sigma = \frac{Median(HH1)}{0.6745} \qquad (8)$$

Input: Noisy image; **Output:** Denoised image

i.   Add noise to the input image (Gaussian).

ii.   Decompose the noisy image using forward SWT.

iii. Compute the noise variance (σ) from the Diagonal subband based on weighted median.

    a. Calculate weights for the coefficients in the diagonal subband using the classical weight function

$$\mathbf{W}(x, y) = \frac{1}{e^{|HH(x,y)|}}$$

    b. Multiply W (x, y) with HH (x, y) to get weighted diagonal subbandHH1 (x, y).

$$\sigma = \frac{Median(HH1)}{0.6745}$$

    c. Compute noise variance from HH1 (x, y) as

iv.   Threshold the wavelet coefficients in detail subbands using modified VisuShrink with golden ratio as

$$T = \sigma(\sqrt{1.618 * Log(N)})$$

v.   Reconstruct using ISWT to get noise free iris image.

vi.   Evaluate the performance using image quality metrics such as MSE, RMSE, PSNR and SNR.

Algorithm 1: Denoising procedure of modified VisuShrink using Golden ratio and weighted median

After computing the noise variance, the new modified VisuShrink is applied to the noisy iris image and the procedure is presented in Algorithm 1.

4   Experimental Results and Discussion

The performance of the proposed method is compared with Median filter, Wiener filter and traditional Visushrink [5]. The proposed method has been implemented in MATLAB. Furthermore, wavelet filters such as Daubechies, Coiflet, Symlet with different lengths are used to evaluate the performance of the proposed method. Initially Gaussian noise is added to the pixels of input iris images at different noise levels. Then the noisy image is decomposed, thresholded and reconstructed to get the noise free image.

4.1   Dataset

The proposed method is tested on Version1 of CASIA database maintained by Chinese Academy of Sciences Institute of Automation [14]. It consists of 756 images collected from 108 persons. The images are resized to 256×256 for implementation and further analysis.

4.2   Quantative Measure

The image quality metrics such as Mean Square Error (MSE), Root Mean Square Error (RMSE), Signal to Noise Ratio (SNR) and Peak Signal to Noise Ratio (PSNR) [6] are used to evaluate the performance. The metrics are shown in Table 1.

Table 1.   Quantitative Metrics

Metric	Formula
MSE	$\dfrac{\sum_{m,n}[I_1(m,n)-I_2(m,n)]^2}{m*n}$
RMSE	$\sqrt{\dfrac{\sum_{m,n}[I_1(m,n)-I_2(m,n)]^2}{m*n}}$
SNR	$10\log_{10}\left(\dfrac{\mathrm{Var}(I_1)}{\mathrm{Var}(I_2)}\right)$
PSNR	$10\log_{10}\left(\dfrac{R^2}{\mathrm{MSE}}\right)$

where I1 is the input image, I2 is the denoised image, m and n are the number of rows and columns in the image respectively, and R is the maximum fluctuation in the input image data type. In this experiment R is set as 255, since the image data type is 8-bit unsigned integer.

The resultant images after applying denoising methods are shown in Fig. 4. The following tables show the mean value of MSE, RMSE, PSNR and SNR of the version1 of CASIA database.

The performance results in spatial domain using median filter and Wiener filter for Gaussian noise are given in Tables 2 and 3 for level 0.001 and 0.003 respectively. The MSE, RMSE, PSNR and SNR of Wiener filter are improved than the Median filter.

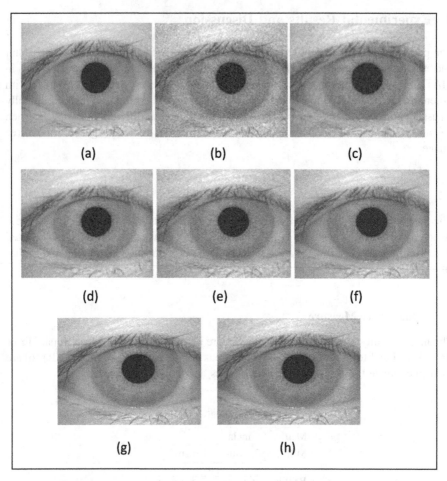

Fig. 4. Denoising using different methods: (a) input iris image; (b) Gaussian noise added an iris image; (c) Denoising using Median filter; (d) Denoising using a Wiener filter; (e) Denoising using Bayeshrink (f) Denoising using Oracleshrink (g) Denoising using VisuShrink; (h) Denoising using the proposed threshold.

Table 2. Median and Wiener filter (0.001)

S.No	Filter	MSE	RMSE	PSNR	SNR
1	MEDIAN	85.90653	9.261019	28.83866	0.020196
2	WIENER	69.81932	8.353088	29.73067	0.065953

Table 3. Median and wiener filter (0.003)

S.No	Filter	MSE	RMSE	PSNR	SNR
1	MEDIAN	205.0414	14.31557	25.05087	-0.01236
2	WIENER	190.6319	13.94519	26.92406	0.050575

The performance results of Bayeshrink and Oracleshrink with Gaussian noise of level as 0.001 based on hard thresholding are given in Tables 4 and 5 respectively.

Table 4. Bayeshrink With Hard Thresholding (0.001)

S.No	Filter	MSE	RMSE	PSNR	SNR
1	COIF2	45.2383	6.7259	31.5757	0.027361
2	COIF4	46.7274	6.8357	31.4351	0.022403
3	SYM2	42.3676	6.509	31.8605	0.039486
4	SYM4	17.111	4.1365	35.7981	0.0054077
5	DB1	15.8223	3.9777	36.1381	0.00449
6	DB2	19.7185	4.4406	35.1821	0.05431

Table 5. Oracleshrink With Hard Thresholding (0.001)

S.No	Filter	MSE	RMSE	PSNR	SNR
1	COIF2	5.436943	2.249841	41.41907	0.198349
2	COIF4	2.713885	1.647275	43.83006	0.133168
3	SYM2	2.796191	1.672151	43.69945	0.002904
4	SYM4	2.645334	1.626365	43.94089	0.140314
5	DB1	1.662774	1.285774	46.00799	0.127279
6	DB2	1.643094	1.277638	46.06698	0.126409

The performance results of traditional universal threshold with Gaussian noise of level as 0.001 based on hard and soft thresholding are given in Tables 6 and 7 respectively.

Table 6. Visushrink with Hard thresholding (0.001)

S. No	Filter	MSE	RMSE	PSNR	SNR
1	COIF2	57.06998	7.551300	30.60800	0.01752
2	COIF4	58.5482	7.648352	30.49722	0.015024
3	SYM2	55.66843	7.457904	30.71622	0.02498
4	SYM4	56.94612	7.543042	30.61758	0.017963
5	DB1	54.70183	7.392654	30.89283	0.04471
6	DB2	55.65834	7.457234	30.71699	0.025131

Table 7. Visushrink With Soft Thresholding (0.001)

S.No	Filter	MSE	RMSE	PSNR	SNR
1	COIF2	60.29405	7.759888	30.37331	0.022128
2	COIF4	61.7512	7.852934	30.26995	0.017728
3	SYM2	58.93478	7.6719	30.47238	0.032777
4	SYM4	60.15211	7.750768	30.38351	0.022236
5	DB1	58.59818	7.649401	30.69854	0.071878
6	DB2	58.97027	7.674268	30.46964	0.032799

The performance results of traditional universal thresholdwith Gaussian noise based on hard and soft thresholding are given in Tables 8 and 9 respectively for noise level 0.003. The MSE, RMSE, PSNR and SNR of traditional universal threshold is improved than median and wiener filter. Among the wavelet filters db1 gives the best result for hard and soft thresholding.

Table 8. Visushrink With hard Thresholding (0.003)

S.No	Filter	MSE	RMSE	PSNR	SNR
1	COIF2	146.4457	12.0984	26.51246	0.019448
2	COIF4	149.7367	12.2336	26.41592	0.015839
3	SYM2	142.6349	11.93999	26.62691	0.029918
4	SYM4	146.1333	12.0855	26.52172	0.019591
5	DB1	139.6193	11.81285	26.72008	0.065331
6	DB2	142.6209	11.93936	26.62741	0.029426

Table 9. Visushrink With soft Thresholding (0.003)

S.No	Filter	MSE	RMSE	PSNR	SNR
1	COIF2	147.9345	12.15956	26.4688	0.01925
2	COIF4	151.1354	12.29038	26.37586	0.014746
3	SYM2	144.223	12.00606	26.57914	0.031774
4	SYM4	147.6758	12.1489	26.47642	0.020021
5	DB1	141.7559	11.90267	26.65444	0.074426
6	DB2	144.2592	12.00753	26.57811	0.031344

The performance results of proposed modified universal threshold based on GR and weighted median with Gaussian noise of level as 0.001 based on hard and soft thresholding are given in Tables 10 and 11 respectively. The MSE, RMSE, PSNR and SNR of proposed modified universal threshold is improved than the traditional universal threshold.

Table 10. Proposed Modified Universal Threshold Based On GR And Weighted Median With Hard Thresholding (0.001)

S.No	Filter	MSE	RMSE	PSNR	SNR
1	COIF2	0.160649	0.400765	56.10798	0.001261
2	COIF4	0.162147	0.402628	56.06775	0.001265
3	SYM2	0.159486	0.399313	56.13946	0.001243
4	SYM4	0.162003	0.402447	56.07169	0.001268
5	DB1	0.13210	0.389936	56.84636	0.000708
6	DB2	0.159571	0.399419	56.13720	0.001224

The performance results of proposed modified universal threshold based on GR and weighted median with Gaussian noise of level as 0.003 based on hard and soft thresholding are given in Tables 12 and 13 respectively. The MSE, RMSE, PSNR and

Table 11. Proposed Modified Universal Threshold Based On GR And Weighted Median With Soft Thresholding (0.001)

S.No	Filter	MSE	RMSE	PSNR	SNR
1	COIF2	2.75806	1.660724	43.75894	0.002775
2	COIF4	2.834323	1.683532	43.64044	0.002622
3	SYM2	2.642567	1.62557	43.94483	0.003462
4	SYM4	2.752286	1.658986	43.76803	0.002752
5	DB1	2.279561	1.509785	44.58671	0.005984
6	DB2	2.642409	1.625521	43.94510	0.003454

Table 12. Proposed Modified Universal Threshold Based On GR And Weighted Median With Hard Thresholding (0.003)

S.No	Filter	MSE	RMSE	PSNR	SNR
1	COIF2	0.091358	0.302224	58.55912	0.001069
2	COIF4	0.090607	0.300975	58.59521	0.001079
3	SYM2	0.09382	0.306275	58.44329	0.001144
4	SYM4	0.093023	0.304965	58.48069	0.001118
5	DB1	0.090517	0.308756	58.95870	0.000419
6	DB2	0.093685	0.306053	58.44965	0.00112

Table 13. Proposed Modified Universal Threshold Based On GR And Weighted Median With Soft Thresholding (0.003)

S.No	Filter	MSE	RMSE	PSNR	SNR
1	COIF2	2.900557	1.703083	43.54018	0.002292
2	COIF4	2.969625	1.723246	43.43792	0.002354
3	SYM2	2.79619	1.672151	43.69945	0.002904
4	SYM4	2.897158	1.702087	43.54525	0.002315
5	DB1	2.331392	1.526874	44.48883	0.004934
6	DB2	2.798167	1.672743	43.69636	0.002913

SNR of proposed modified universal threshold is improved than median, wiener and the traditional universal threshold.

In summary, the performance of proposed modified universal threshold based on GR and weighted median with hard thresholding is improved than Median filter, Wiener filter, Bayeshrink, Oracleshrink and traditional universal threshold for iris image denoising. Hard thresholding provides better results when compared to soft thresholding. From the above tables, it is clearly understood that among all the wavelet filters db1 outperforms. In Fig. 5, the PSNR of various denoising methods is compared with the proposed method. Figures 4 and 5 shows that the performance of the proposed is better than that of existing methods for iris denoising.

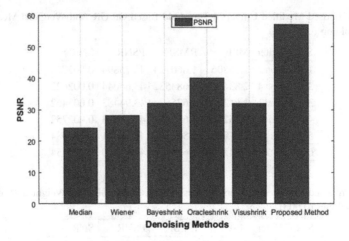

Fig. 5. Comparison of PSNR of different denoising methods

5 Conclusion

A novel method for denoising iris image using undecimated wavelet transform and a threshold based on weighted median is proposed in this paper. Initially Gaussian noise is added to the iris image at different noise levels and SWT is used to decompose the noisy iris image into four subbands. Then the new modified universal threshold is applied to the wavelet coefficients using hard and soft thresholding. Finally the noise free iris image is reconstructed from the thresholded subbands using inverse SWT. The daubechies wavelet filter at level 2 gives the best result than symlet and daubechies filters. The proposed method outperforms Median, Wiener, Bayeshrink, Oracleshrink and traditional universal threshold in many denoising applications. The quantitative measures show that the new modified universal threshold removes Gaussian noise present in iris image more effectively. The proposed threshold is simple in computation and yet it is effective in denoising.

Acknowledgements. The first and second author would like to thank UGC, New Delhi for the financial support received under UGC Major Research Project No. 43-274/2014(SR).

References

1. Sutcu, Y., Tabassi, E., Sencar, H.T., Memon, N.: What is biometric information and how to measure it?. In: IEEE International Conference on Technologies for Homeland Security (HST), pp. 12–14 (2013)
2. Sasirekha, K., Thangavel, K.: A comparative analysis on fingerprint binarization techniques. Int. J. Comput. Intell. Inform. 4(3), 163–168 (2014)

3. Sasirekha, K., Thangavel, K.: A novel feature extraction algorithm from fingerprint image in wavelet domain. In: Senthilkumar, M., Ramasamy, V., Sheen, S., Veeramani, C., Bonato, A., Batten, L. (eds.) Intelligence, Cyber Security and Computational Models, ICC3 2015, Advances in Intelligent Systems and Computing, vol. 412, pp. 135–143. Springer, Heidelberg (2016)
4. Sanjay, N., Shrivastava, T.A., Upadhyay, A.R.: Advanced denoising technique for Iris images. In: International Conference on Systemics, Cybernetics and Informatics, vol. 3, pp. 106–109 (2011)
5. Sasirekha, K., Thangavel, K.: A novel wavelet based thresholding for denoising fingerprint image. In: IEEE International Conference on Electronics, Communication and Computational Engineering, pp. 119–124 (2014)
6. Mohideen, S.K., Perumal, S.A., Krishnan, N., Sathik, M.M., Kumar, T.C.R.: Image denoising multi-wavelet and threshold. In: IEEE International Conference on Computing, Communication and Networking, pp. 1– 5 (2008)
7. Donoho, D.L., Johnstone, I.M.: Ideal spatial adaptation via wavelet shrinkage. Biometrika **81**, 425–455 (1994)
8. Donoho, David L., Johnstone, I.M.: Adapting to unknown smoothness via wavelet shrinkage. J. Am. Stat. Assoc. **90**(432), 1200–1224 (1995)
9. Grace Chang, S., Yu, B., Vetterli, M.: Adaptive wavelet thresholding for image denoising and compression. IEEE Int. Trans. Image Process. **9**(9), 1532–1546 (2000)
10. Wavelet_Denoising. www.mors.org/UserFiles/file/.../Wavelet_Denoising.pdf
11. Stakhov, A.P.: The generalized principle of the golden section and its applications in mathematics, science and engineering. Chaos, Solutions Fractals **26**(2), 263–289 (2005)
12. Hashemiparast, S.M., Hashemiparast, O.: Multi Parameters Golden Ratio and Some Applications. Appl. Math. **2**(7), 808–815 (2011)
13. Wang, C.-Y., Li, L.-L., Yang, F.-P., Gong, H.: A new kind of adaptive weighted median filter algorithm. In: IEEE International Conference on Computer Application and System Modeling, vol. 11, pp. 667–671 (2010)
14. Iris Database: CASIA-IrisV1. http://biometrics.idealtest.org

Analogy Removal Stemmer Algorithm
for Tamil Text Corpora

M. Thangarasu[✉] and H. Hannah Inbarani

Department of Computer Science, Periyar University, Salem 636011, India
thangarasumathan@gmail.com, hhina@gmail.com

Abstract. Stemming is the process of generating root word from the given inflectional word. Tamil Language has technical challenges in stemming because it has rich morphological patterns than other languages, so Analogy Removal Stemmer (ARS) is proposed in this research, to find stem word for the given inflection Tamil word from text corpora. The performance of the proposed approach is compared with Light Stemmer (LS) and Improved Light Stemmer (ILS) algorithms based on correctly and incorrectly predicted stem words. The experimental result clearly shows that the proposed approach ARS for Tamil corpora performs better than the LS and ILS algorithm.

Keywords: Natural language processing · Tamil stemmer · Tamil morphology · Lexical analysis

1 Introduction

Word lists are required resources in many disciplines, from language learning to morphology. Word list is developed usually from a corpus [3, 9–11]. Stemming is a technique to transform different inflections and derivations of the same word to one common stem. Stem can mean both prefix and suffix removal from the given input words. Stemming can, for example, be used to ensure that the greatest number of relevant matches is included in search results. A word's stem is its root or basic form: for example, the stem of a plural noun men, ANkaL (ஆண்கள்) is the singularman, (AN (ஆண்)), likewise the stem of a past-tense verb ((Acted), nadiththEn (நடித்தேன்) is the present tense ((act), nadi (நடி)). The stem is yet not to be confused with a word lemma; the stem does not have to be a definite word itself. Instead the stem can be assumed to be the least familiar denominator for the morphological variants.

In this paper, significant effort is made to generate word (or verb) from the giving inflectional word. Tamil is a Dravidian and regional language of Tamil Nadu, India. It has a huge number of morphological variants for a word. Approximately a single verb of Tamil language has more than 3000 morphological forms so this is the real challenge to get a verb from the inflectional Tamil word [2, 29]. Word frequency can have several viewpoints for computational linguistics or information theory; this is called unigram list and can be seen as a solid representation of a corpus [4, 12, 13, 15]. Unigram is used to identify the lexical stem which consists of an identification of a string of letters which co-occurs in a large corpus with various distinct suffixes [5, 16, 17, 24].

© Springer Nature Singapore Pte Ltd. 2016
S. Subramanian et al. (Eds.): CSI 2016, CCIS 679, pp. 70–81, 2016.
DOI: 10.1007/978-981-10-3274-5_6

The purpose of this paper is to establish a wide range of stems and suffix possibilities as possible by giving corpus from a natural language Tamil.

2 Related Work

Earlier, Stemmer was primarily developed for English [1], but later due to the corpus growth of languages other than English, there was a bigger demand from the research community to develop stemmers for other languages too. Adam Kilgarriff and et al., proposed monolingual and bilingual word lists for language learning, using corpus methods, for nine languages and thirty-six language pairs [4]. StelaManova proposed Suffix combinations in Bulgarian pars ability and hierarchy-based ordering for getting affix order and pars ability Hypothesis for Bulgarian language [6]. In the year of 2013, Noam Faust proposed decomposing the feminine suffixes of Modern Hebrew: amorpho-syntactic analysis for getting root word from the distributed morphology [7]. Laurie Bauer proposed a method for large corpora that may help determine the output of a variable rule in morphology where the productivity of the process is involved. If that is the case, the notion of productivity has to be re-evaluated [8]. In Indian languages, the initial work was reported by Ramanathan and Rao [18, 27] to perform longest match stripping for building a Hindi stemmer. In 2001, Shambhavi et al. introduced Kannada morphology analyzer and generator using tire [19]. Zahurul.MD et al. proposed a lightweight stemmer for Bengali [20] in the year of 2009 for developing the Bengali language spell checker. In the year of 2009, Assas-band an affix-exception list based Urdu stemmer [21] was developed by Qurat-Ul-AinAkram and et al. and in this work lexical lookup method (Assas-band) is used to stem the Urdu inflectional words from Urdu root word (Verb).

In 2010, Dinesh Kumar and Prince Rana proposed the design and development of stemmer for Punjabi [22], it uses Brute Force algorithm for stem the Punjabi words. In the year of 2010, Vijay Sundar et al. introduced Malayalam stemmer for information retrieval [23] in this research the Finite State Automata (FSA) method is used to stem the Malayalam words. Tamil morphological analyzer [18] proposed by Vijay Sundar et al. in the year of 2010. In this research matching is performed based on the given input for example a Tamil verb have more number of forms so the forms of the verb stored in look-up table. So the algorithm match the forms of input word from look-up table if match found the result displayed else the stemmer shows notification or intimation.

3 Stemmers for Tamil Language

Rule for Plural to singular conversion. Supposing the given inflectional word is I, the length of the I is represented as Il. If Il ends with plural form "கள் (kaL)" then Il-2 is applied to truncate the last two characters of the string and the result is stored in the new string as In and the following rule is used to convert the plural to singular form.

In = {if I ends with "கள் (kaL)" then Il -2 truncates the last two characters of the string} e.g. அவர்கள் [They], I = அவர்கள் [They], Length of the I is 5, so Il = 5. In the next iteration "I" checks with rule, ends with "கள்". In the next iteration I = Il -2, and finally we get new string In = அவர். Likewise Tamil language has many rules for plural to singular conversions (களுக்கு, etc.) and this research work includes different types of plural to singular conversion rules.

For example His/hobby/is/reading/books [puththakangaL/padippathu/avarathu/pozhuthuPOkku(புத்தகங்கள்/ படிப்பது /அவரது /பொழுதுபோக்கு)] Using the plural to singular conversion rule, the above given sentence can be converted into following form puththakangaL (புத்தகங்கள்)/padippathu(படிப்பது)/avarathu(அவரது) / pozhuthuPOkku(பொழுதுபோக்கு).

His/hobby/is/reading/book, puththakam(புத்தகம்)/padippathu(படிப்பது)/avarathu (அவரது) /pozhuthuPOkku(பொழுதுபோக்கு), From the sentence the word puththakangaL (புத்தகங்கள்) changed into puththakam (புத்தகம்) using the plural to singular rule.

Rules for Noun, Verb and Adjective Truncation. Tamil Language Nouns. Tamil has a widespread case system. Root nouns can assume eight different morphological shapes depending on their role in a sentence. Singular and plural forms are also distinguished through inflections. Suffixes are attached to stem word of the noun. Tables 1, 2, 3 show the rules for noun, compound noun, verb and infinite verb truncation for given inflectional word. These rules are used to design stemmer algorithm for Tamil words. Existing algorithm and proposed algorithm uses the following execution steps which includes plural to singular conversion.

I -(Inflectional word), Il -(Length of the Inflectional word), Il-r -(Truncation based on the rules). Consider the example, an inflectional word word I is நடித்தான் (He acted), Il is 5, according to the Table 2, third singular male rule is matched with I based on the rule and truncation can be done using the algorithm and finally we get the stemmed word நடி (He act).

Tamil Language Compound Nouns. A noun also occurs in different compound forms as well. It can be made up of numerous units where each unit expresses an exact grammatical meaning. The Tamil noun, "Odikondurunthavanai (ஓடிகொண்டிருந்தவனை)", which translates as, "the male who was running", provides information on tense, number, gender, person and case. This noun is actually obtained from the full non-infinite verb, "Odikkondu (ஓடிக்கொண்டு)", which means, "running". In English, deriving nouns from verbs is seen too. The full finite verb, "run", for instance, could be changed into a noun by adding the suffix "er" to its stem, so that it becomes "runner". But, while Tamil is an agglutinating language, English is not.

Tamil Language Verbs. Tamil verbs may be major or auxiliary. They also exist in finite and non-finite forms just as in English. Tamil finite verbs, however provides much more grammatical information than English finite verbs do, in that they mark number, person, gender, case, tense, mood, etc. In Tables 2 and 3 below, one can observe the different finite and non-finite morphological constructions for the verb "nadi(நடி)" [act].

Table 1. Font Noun in Tamil language.

Singular	AN (ஆண)/ man	seti (செடி) /plant
Oblique stem	AN (ஆண)	seti(செடி)-
Nominative stem	AN (ஆண)	seti(செடி)
Accusative stem	AN (ஆண)-ai(ஐ)	seti(செடி)-ai(ஐ)
Dative stem	AN (ஆண)-ukku(உக்கு)	seti(செடி)-ukku(உக்கு)
Sociative stem	AN (ஆண)-odu(ஒடு)	seti(செடி) -odu(ஒடு)
Genitive stem	AN (ஆண)-udaiya(உடைய)	seti(செடி)-udaiya(உடைய)
Instrumental stem	AN (ஆண)–aal(ஆல்)	seti(செடி) -aal(ஆல்)
Locative stem	AN (ஆண)-idam(இடம்)	seti(செடி)-marath(இல்)
Ablative stem	AN(ஆண)–idamirunthu(இட மிருந்து)	seti(செடி) -ilirunthu(இலிருந்து)
Vocative stem	AN (ஆண)-e(எ)	seti(செடி) -e(எ)
Plural	**ANkaL (ஆண்கள்)/ men**	**setikaL (செடிகள்)/ plants**
Oblique stem	ANkaL (ஆண்கள்)-	setikaL(செடிகள்)-
Nominative stem	ANkaL (ஆண்கள்)	setikaL(செடிகள்)
Accusative stem	ANkaL (ஆண்கள்)-ai(ஐ)	setikaL(செடிகள்)-ai(ஐ)
Dative stem	ANkaL(ஆண்கள்)-ukku(உக்கு)	setikaL(செடிகள்)-ukku(உக்கு)
Sociative stem	ANkaL (ஆண்கள்)-odu(ஒடு)	setikaL(செடிகள்)-odu(ஒடு)
Genitive stem	ANkaL(ஆண்கள்)-udaiya(உடைய)	setikaL(செடிகள்)-udaiya(உடைய)
Instrumental stem	ANkaL (ஆண்கள்)-aal(ஆல்)	setikaL(செடிகள்)-aal(ஆல்)
Locative stem	ANkaL(ஆண்கள்)-idam(இடம்)	setikaL(செடிகள்)-il(இல்)
Ablative stem	ANkaL(ஆண்கள்)-idamirunthu(இடமிருந்து)	setikaL(செடிகள்)-ilirunthu(இலிருந்து)
Vocative stem	ANkaL (ஆண்கள்)-e(எ)	setikaL(செடிகள்)-e(எ)

In this research stemming algorithm uses the two major technique called as

i. Single Prefix non-recursively (SP)
ii. Suffix Prefix Suffix (SPS)

The following rules show the difference between the SP and SPS rules

Rule for Morphological Analysis for Single Prefix non-recursively (SP)

Notes = note + Noun + PluraL

அவர்கள்=அவர்+கள்

Rule for Morphological Analysis for via Suffix Prefix Suffix (SPS)

Playing = play + Verb + Continuous

Table 2. Verbs in Tamil language.

	Past	Present	Future	Future-Neg
I singular	nadi (நடி)-ththEn(த்தே தன்)	nadi(நடி)-kkiREn(க்கிறேன்)	nadi(நடி)-ppEn(ப்பேன்)	nadi(நடி)-kkamaattEn(க்கமாட் டேன்)
II singular	nadi(நடி)-thhAi(த்தாய்)	nadi(நடி)-kkiRaai(க்கிறாய்)	nadi(நடி)-ppaai(ப்பாய்)	nadi(நடி)-kkamataai(க் கமடாய்)
III singular male	nadi(நடி)-thhaan(த்த ான்)	nadi(நடி)-kkiRaan(க்கிறான்)	nadi(நடி)-ppaan(ப்பான்)	nadi(நடி)-kkamaattaan(க்கமாட்டான்)
III singular female	nadi(நடி)-thhaaL(த்த ாள்)	nadi(நடி)-kkiRaaL(க்கிறாள்)	nadi(நடி)-papal(ப்பாள்)	nadi(நடி)-kkamaattaaL(க்கமாட்டாள்)
III singular hon	nadi(நடி)-thhaar(த்தார்)	nadi(நடி)-kkiRaar(க்கிறார்)	nadi(நடி)-ppaar(ப்பார்)	nadi(நடி)-kkamaatdaar(க்கமாட்டார்)
III singular inan	nadi(நடி)-thhathu(த் தது)	nadi(நடி)-kkiRathu(க்கிறது)	nadi(நடி)-kkum(க்கும்)	nadi(நடி)-kkaathu(க் காது)
I plural	nadi(நடி)-thOm(ட்தே ாம்)	nadi(நடி)-kkiROm(க்கிறோம்)	nadi(நடி)-pPOm(ப்போம்)	nadi(நடி)-kkamaatTOm க்கமாட்டோாம்)
II plural	nadi(நடி)-thhIrkaL(த் தீர்கள்)	nadi(நடி)-kkiRIrkaL(க்கிறீர்க ள்)	nadi(நடி)-ppIRkaL(ப்பீற்கள்)	nadi-(நடி)-kkamaattaarkaL(க்கமா ட்டீர்கள்)
III plural an	nadi(நடி)-thhaarkaL(த்தார்கள்)	nadi(நடி)-kkiRaarkaL(க்கிறார் கள்)	nadi(நடி)-ppaarKaL(ப்பார்கள்)	nadi(நடி)-kkamaattaarkaL(க்கமாட்டார்கள்)
III plural inan	nadi(நடி)-thhthana(த் தன)	nadi(நடி)-kkinRana(க்கின்றன)	nadi(நடி)-kkum(க்கும்)	nadi(நடி)-kkaathu(க் காது)

பாடுகிறார்கள்=பாடு+கிறார்+கள். The major difference between SP and SPS is SP truncates the possible suffixes for the given input and sometimes it's given infinite verb as output. So in SPS the more rules are used (detailed in Tables 1, 2, 3) for getting the finite verb (root word) for giving input. The existing and proposed algorithms are detailed in the following sequel.

Table 3. Non-finite verbs in Tamil language.

Conjunctive	nadi(நடி)-thu (து)
Infinitive	nadi(நடி)-kka(க்க)
Neg.verbal participle	nadi(நடி)-kkaamal(க்காமல்)
Conditional	nadi(நடி)-thaal(தால்)
Neg. conditional	nadi(நடி)-kkaanittaal(க்கனிட்டல்)
Neg.relative participle	nadi(நடி)-kkaatha(க்காத)
Neg. verbal noun	nadi(நடி)-kkaathathu (க்காதது)
Deverbal nouns	nadi-(நடி)thal(தால்);nadi(நடி)-ppu(ப்பு); nadi(நடி)-kkai(க்கை)

3.1 Light Stemmer (LS)

Light stemmer is the kind of the rule based stemmer. It works by truncating all possible suffixes from the given inflectional word (removes suffixes recursively and a single prefix non-recursively that is called SP). Light stemming is used to find the representative indexing form of giving word by the application of truncation of suffixes [14, 21, 25, 26]. The objective of light stemmer is to protect the word meaning intact and increase the retrieval performance of an IR system. A Light Stemmer Algorithm for Tamil word is projected in Fig. 1 [30].

Light Stemmer Algorithm
Input   :   List of Tamil words
Output : Stemmed(Root) words

Step1: remove the total composite plural. E.g(அவர்கள், செல்கிறார்கள், வந்தார்கள்...)
அவர்கள்= அவர், செல்கிறார்கள்= செல்கிறார், வந்தார்கள்= வந்தார்

Step2: remove the frequencies from the word suffixes. Eg(அவர், செல்கிறார், வந்தார்)அவர்= அவர், செல்கிறார்= செல்கி, வந்தார்= வந்த

A : remove றார்
செல்கிறார்=செல்கி

B : removeஎர்
வந்தார்= வந்த

Step 3: According to the recognized suffix, the subsequently probable suffix list is produced using rules mentioned in the tables.

Fig. 1. Light stemmer algorithm

3.2   Improved Light Stemmer (ILS)

Improved Light Stemming is also an algorithm which removes suffixes recursively and a single prefix non-recursively which is called as SP. The similar defined affixation terms list were used but modified using Suffix-Prefix-Suffix truncating process called SPS. Notice that, some of Tamil words use (Thiru, Thirumathi) prefix as a declarative term (e.g., Thiru. Dr. A.P.J. Abdul Kalam). Therefore to classify the words based on the rules like Without-Thiru (WOTH the stemmer accepts the non-stemmed words after removing the prefixed Thiru) and With Thiru (WTH stemmer acquires the whole non-stemmed word) an algorithm is proposed in [30]. This work applies improved light stemming concept to replace plural terms, adjectives and tense words [30]. The Improved Light Stemmer Algorithms projected in Fig. 2 [28].

Improved Light stemmer Algorithm
Input: List of   Tamil   words
Output: Stemmed(Root) words

Step 1:   Remove the plural forms from the inflectional words.

Step 2:   From the step 1, plural word is converted into singular word, The word is also checked for adjective using SPS; if it is found, then its equivalent verb is substituted. Example, the term 'diya(டிய)' in Odiya(ஓடிய) will be changed to 'Oodu(ஓடு)'.

Step3:   After the adjectives are converted to main word, the tenses are   eliminated   so   that Paadiya(பாடிய),Paadukinra(பாடுகின்ற) and Paadum(பாடும்) will be changed to Paadu(பாடு).

Step 4:   According to the recognized suffix, the subsequently probable suffix list is produced using adjective and tense elimination rules.

Fig. 2. Improved Light Stemmer Algorithm

3.3   Analogy Removal Stemmer (ARS)

Light stemmer uses the SP technique so it truncates all possible suffixes from the inflectional word. Sometimes it gives infinite verb from the LS word செல்கிறார்கள் to give the infinite verb செல்கி. This is the major problem in LS so Improved Light Stemmer algorithm rectify this problem using Suffix Prefix Suffix truncating process called SPS. Sometimes ILS gives infinite verb during suffix truncation so we need to rectify the issue using affix truncation. The FBARS uses the Suffix Prefix Affix rules for generating root word from the inflectional word. Consider the following rule to rectify the SPS problem. Figure 3 is represented as the FBARS algorithm.

Rule for Morphological Analysis for via Suffix Prefix Affix (SPA)

(i) Cheeriest = cheer + Adj + Superlative
(ii) Got = get + Verb + Past
பைந்தளிர்(painthaLir)=பசுமை(pasumai)+ தளிர் (thaLir)

Frequency Based Analogy Removal Stemmer
Input: List of   Tamil   words
Output: Stemmed(Root) words

Step 1:   Remove the plural forms in the inflectional words.
Step 2:   From the step 1, plural word is converted into singular word,
          The word is also checked for adjective using SPA Rules; if it
          is found, then its equivalent verb is substituted. Example,
          the term 'diya(டிய)' in Odiya(ஓடிய) will be changed to
          'Oodu(ஓடு)'.

Step3:   After the adjectives are converted to main word, the tenses
          are   eliminated   so   that   Paadiya(பாடிய),
          Paadukinra(பாடுகின்ற)   and   Paadum(பாடும்)   will   be
          changed to Paadu(பாடு).
Step 4:   According to the recognized suffix and affix, the
          subsequently probable suffix and affix list is produced
          using adjective and tense elimination rules.

Fig. 3. Analogy Removal Stemmer Algorithm

4   Experimental Results and Analysis

Tamil corpus collected from Central Institute of Indian Languages (CILL) is given as
input to this process. The corpus consists of five documents called as Dataset I to
Dataset V. Dataset I consists of 2497 and it has 2224 unique words. A unique word is
used to calculate an exact accuracy of the stemmer algorithm, because total number of
words has repeated words (redundancy). It may affect the stemmer accuracy so we
generate the unique words. Likewise we identify unique words from Dataset II to
Dataset V for calculating stemmer accuracy. The details are tabulated in Table 4. The
goal of stemmer algorithm is to achieve highest accuracy for the given dataset and the
stemmer performance is evaluated by using the interrelated measures of precision and
recall. Precision is defined as the ratio of the number of words stemmed to total number
of unique words. Recall is defined as the ratio of the number of words stemmed
correctly to the total number of words stemmed. The precision (P) formula is provided
in Eq. 1 and Recall (R) is given in Eq. 2. Equations 3 and 4 represent Accuracy
(A) and F-Measure (F).

$$Precision(P) = (Number\ of\ words\ stemmed/Total\ number\ of\ unique\ words) * 100 \tag{1}$$

$$Recall(R) = (Number\ of\ words\ stemmed\ correctly/\ number\ of\ words\ stemmed) * 100 \tag{2}$$

$$Accuracy(A) = (Number\ of\ words\ stemmed\ correctly/Total\ unique\ words) * 100 \tag{3}$$

$$F-\text{Measure}(F) = (2PR/P + R) \tag{4}$$

Table 4 shows an overview of the characteristics of the trained corpus. Figure 4 compares stemming algorithms based on the number of words stemmed correctly from Table 4 and Fig. 4. The proposed ARS algorithm gives more stemmed words compared to LS and ILS.

The comparison of stemmer algorithms based on Accuracy, Precision, Recall and F-Measure is shown in Table 5.

Table 4. Test dataset.

Dataset	No of words	No of unique words	No of words stemmed			No of words stemmed correctly		
			LS	ILS	ARS	LS	ILS	ARS
Dataset I	2497	2224	2047	2106	2123	2033	2098	2118
Dataset II	5078	4289	4081	4137	4194	4067	4128	4184
Dataset III	1946	1704	1649	1682	1695	1643	1677	1686
Dataset IV	6071	5167	4958	4969	5017	4943	4965	5002
Dataset V	5218	4656	4526	4601	4613	4512	4597	4602

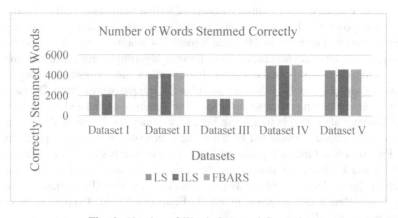

Fig. 4. Number of Words Stemmed Correctly

Table 5. Test dataset (number represented in %).

Test Data	Accuracy			Precision			Recall			F-measure		
	LS	ILS	ARS	LS	ILS	ARS	LS	ILS	ARS	LS	ILS	ARS
Dataset I	91.4	94.3	95.2	92.0	94.6	95.4	99.3	99.6	99.7	95.5	97.0	97.5
Dataset II	94.8	96.2	97.5	95.1	96.4	97.7	99.6	99.7	99.7	97.2	98.0	98.6
Dataset III	96.4	98.4	98.9	96.7	98.7	99.4	99.6	99.7	99.4	98.1	99.1	99.4
Dataset IV	95.6	96.0	96.8	95.9	96.1	97.0	99.6	99.9	99.7	97.7	97.9	98.3
Dataset V	96.9	98.7	98.8	97.2	98.8	99.0	99.6	99.9	99.7	98.3	99.3	99.3

Table 6. Average results (number represented in %).

Accuracy			Precision			Recall			F-measure		
LS	ILS	ARS	LS	ILS	ARS	LS	ILS	ARS	LS	ILS	ARS
95.02	96.72	97.44	95.38	96.92	97.7	99.54	99.74	99.64	97.36	98.26	98.62

The comparison of average of experimental results is shown in Table 6. The Average formula is described in Eq. 5.

$$Average(A) = (Sum\ of\ observations/Number\ of\ observation) * 100 \qquad (5)$$

Table 6 describes that average maximum accuracy achieved by ARS which is 97.44% and F-score which is 98.62%. Based on Experimental analysis, in terms of accuracy, precision and F-score, proposed stemmer performs better than existing algorithms LS and ILS. The result of the average experimental results is shown in Table 6. Proposed stemmer is unsupervised and language independent. Corpus is used to derive set of powerful suffixes and it does not need any linguistic knowledge. Hence, this approach can be used for developing stemmers for other languages that are morphologically rich. The performance of the proposed paradigm is evaluated in terms of accuracy, precision, recall and F-score.

5 Conclusion

Stemming plays vital role in Tamil information retrieval, compared with all other languages. The stemming effects are very large, compared to that found in review on other stemming algorithms. According to experimental results, the proposed Analogy Removal Stemmer is performing better than the existing Tamil stemmer algorithms Light stemmer and Improved Light Stemmer. ARS algorithm is robust. It does not require complete sentences and it does not try to handle every single case so ARS algorithm is suitable for Information Retrieval System (IRS) of Tamil language since it performs efficiently for morphological rich language Tamil.

References

1. Porter, M.F.: An algorithm for suffix stripping. Readings Inf. Retrieval **4**, 313–316 (1980)
2. Ramachandran, V.A., Krishnamurthi, I.: An iterative suffix stripping Tamil stemmer. In: Satapathy, S.C., Avadhani, P.S., Abraham, A. (eds.) Proceedings of the InConINDIA 2012. AISC, vol. 132, pp. 583–590. Springer, Heidelberg (2012)
3. Savoy, J.: A stemming procedure and stop word list for general French Corpora. J. Am. Soc. Inf. Sci. **50**, 944–952 (1999). Wiley
4. Kilgarriff, A., Charalabopoulo, F.: Corpus-based vocabulary lists for language learners for nine languages. Lang. Resour. Eval. **48**, 121–163 (2014). Springer
5. Goldsmith, J.A., Higgins, D., Soglasnova, S.: Automatic language-specific stemming in information retrieval. In: Peters, C. (ed.) CLEF 2000. LNCS, vol. 2069, pp. 273–283. Springer, Heidelberg (2001). doi:10.1007/3-540-44645-1_27
6. Manova, S.: Suffix combinations in Bulgarian: parsability and hierarchy-based ordering. Morphology **20**, 267–296 (2010). Springer
7. Faust, N.: Decomposing the feminine suffixes of modern Hebrew: a morpho-syntactic analysis. Morphology **23**, 409–440 (2013). Springer
8. Bauer, L.: Grammaticality, acceptability, possible words and large corpora. Morphology **24**, 83–103 (2014). Springer
9. Esher, L.: Autonomous morphology and extramorphological coherence. Morphology **24**, 325–350 (2014). Springer
10. Jenny, A.: Booij, Geert: the grammar of words: an introduction to linguistic morphology. Morphology **24**, 433–434 (2014). Springer
11. Pertsova, K.: Interaction of morphological and phonological markedness in Russian genitive plural allomorphy. Morphology **25**, 229–266 (2015). Springer
12. Sims, A.D., Parker, J.: Lexical processing and affix ordering: cross-linguistic predictions. Morphology **25**, 143–182 (2015). Springer
13. Andreou, M.: Lexical negation in lexical semantics: the prefixes in and dis. Morphology **25**, 391–410 (2015)
14. Braschler, M., Ripplinger, B.: How effective is stemming and de compounding for German text retrieval. Inf. Retrieval **7**, 291–316 (2004)
15. Larkey, L.S., Ballesteros, L., Connell, M.E.: Improving stemming for Arabic information retrieval: light stemming and co-occurrence analysis. In: SIGIR 2002. ACM (2004)
16. Korenius, T., Laurikkala, J., Järvelin, K., Juhola, M.: Stemming and Lemmatization in the Clustering of Finnish Text Documents, CIKM 2004. ACM (2004)
17. Hollink, V., Kamps, J., Monz, C., de Rijke, M.: Monolingual Document Retrieval for European Languages. Kluwer Academic Publishers, Dordrecht (2003)
18. Ramanathan, A., Rao, D.: A lightweight stemmer for Hindi. In: Proceedings of the 10th Conference of the European Chapter of the Association for Computational Linguistics (EACL) on Computational linguistics for South Asian Language (2003)
19. Shambhavi, B.R., Kumar, P.R.: Kannada morphological analyzer and generator using trie. Int. J. Comput. Sci. Netw. Secur. **11**, 112–116 (2011)
20. Islam, Z., Uddin, M.N., Khan, M.: A light weight stemmer for bengali and its use in spelling checker. In: Proceedings of First International Conference on Digital Communication and Computer Applications (DCCA 2007), pp. 19–23 (2007)
21. Akram, Q.U.A., Naseer, A., Hussain, S.: Assas-band, an affix-exception-list based Urdu stemmer. In: Proceedings of the 7th Workshop on Asian Language Resources, pp. 40–47 (2009)

22. Hybrid Approach for Stemming in Punjabi. Int. J. Comput. Sci. Comput. Netw. http://www. ijcscn.com/Documents/Volumes/vol3issue2/ijcscn2013030206.pdf
23. Ram, V.S., Devi, S.L.: Malayalam stemmer. In: Morphological Analysers and Generators, Mona Parakh, LDC-IL, Mysore, pp. 105–113 (2010)
24. Mudassar, M.: Majgaonker: discovering suffixes: a case study for Marathi language. Int. J. Comput. Sci. Eng. **02**, 2716–2720 (2010)
25. Sasidhar, B., Yohan, P.M.: Named entity recognition in Telugu language using language. Int. J. Comput. Appl. **22**, 30–34 (2011)
26. Ameta, J., Joshi, N., Mathur, I.: A lightweight stemmer for Gujarati. In: 46th Annual National Convention of Computer Society of India. Organized by Computer Society of India Gujarat Chapter. Sponsored by Computer Society of India and Department of Science and Technology, Govt. of Gujarat and IEEE Gujarat Section
27. Mishra, U., Chandra, P.: MAULIK: an effective stemmer for Hindi language. Int. J. Comput. Sci. Eng. **4**, 711–717 (2012)
28. Thangarasu, M., Manavalan, R.: Design and development of stemmer for Tamil language: cluster analysis. Int. J. Adv. Res. Comput. Sci. Softw. Eng. **3**, 813–818 (2013)
29. Thangarasu, M., Manavalan, R.: A literature review: stemming algorithms for Indian languages. Int. J. Comput. Trends Technol. **4**, 2582–2584 (2012)
30. Thangarasu, M., Manavalan, R.: Stemmers for Tamil language: performance analysis. Int. J. Comput. Sci. Eng. Technol. **4**, 902–908 (2012)

Item Refinement for Improved Recommendations

R. Latha[✉] and R. Nadarajan

Department of Applied Mathematics and Computational Sciences,
PSG College of Technology, Coimbatore, India
lathapsg@yahoo.co.in, nadarajan_psg@yahoo.co.in

Abstract. Recommender systems serve as business tools which make use of knowledge discovery techniques to reshape the world of E-Commerce. Collaborative filtering (CF), the most effective type of recommender systems, predicts user preferences by learning from past user-item relationships. Prediction algorithms are based on similarity between item vectors or user profiles. However similarity computations become less efficient if item vectors or user profiles do not contain enough ratings. A technique which is based on Pseudo Relevance Feedback is proposed to expand item vectors in order to make them contain more ratings. The proposed approach first expands item profiles and refines the expansion in order to remove expansion deviations. The experiments on Movie-Lens data set show that the proposed technique is efficient in expanding the rating matrix and outperforms state of the art collaborative filtering techniques for providing more efficient predictions.

1 Introduction

In many online markets, consumers are faced with extremely large volume of products and information from which they can choose [8]. With so many products available on most websites, a customer may get lost. To alleviate this information overload problem, many web sites attempt to help users by incorporating a product recommender system. A product recommendation system provides with a list of items and/or web-pages that are likely to interest them [25] in order to narrow down the selection to the right choice. Schafer et al. [24] explains how a recommender system helps E-Commerce to increase sales. Many online business systems such as Amazon.com^TM (www.amazon.com), CDNOW^TM (www.cdnow.com), eBay.com^TM (www.ebay.com), Levi Straus^TM (www.levis.com), Match Maker (www.moviefinder.com) and Netflix (www.netflix.com) are using recommender systems to provide personalized suggestions.

Broadly speaking, Recommended systems use two common approaches for making recommendations. They are Collaborative Filtering (CF) and Content Based filtering (CB) [3]. Pazzani et al. [15] defines that content based algorithms base their recommendations on the contents of items and profiles of users. The profiles allow programs to associate users with matching products. CB techniques

© Springer Nature Singapore Pte Ltd. 2016
S. Subramanian et al. (Eds.): CSI 2016, CCIS 679, pp. 82–96, 2016.
DOI: 10.1007/978-981-10-3274-5_7

use the assumption that items with similar objective features will be rated similarly and users with similar taste will prefer items in a similar manner [18]. Lops et al. [13] discusses that CB techniques are specific to a domain and the scope of them is limited to the domain for which they are proposed.

CF is the most popular and successful technique for recommendation systems [5]. CF [8] relies only on the past user behavior (ratings, preferences, purchase history, time spent etc.). The most common form of CF approaches is the 'Nearest based approach (called kNN)'. These kNN methods identify pairs of items that tend to be rated similarly or like-minded users with similar history of rating or purchasing, in order to deduce unknown relationships between users and items [16]. Breese et al. [3] researched the ways of improving the prediction accuracy, using Pearson's correlation coefficient, vector similarity, default voting, inverse user frequency, and case amplification. However, the major limitation of CF techniques is that the similarity calculations are based on common items/users and therefore unreliable when data are sparse and the common items/users are very few. Techniques available in the literature which address cold start user/item make use of content properties [20] or user's demographic information [12,26] to provide solutions for cold start problems, but the proposed work considers only the rating profiles to alleviate the problem.

In order to improve the efficiency of a recommender system, a vector refinement technique is proposed in this paper. The proposed technique consists of three steps. In the first step, for each item vector, its neighbouring items are identified and in the second step, each item vector is expanded by including ratings from its neighbouring items. In the third step, the expansion deviations are corrected so that new train is created for prediction computations. This results in item vectors with more number of ratings. Ultimately increased ratings lead to better predictions. Item refinement can be done in offline and so only prediction computations are done online.

The geometric view of item vector expansion is shown in Fig. 1. Each user is considered as a dimension to represent items in vector space. Therefore the number of dimensions is equal to number of users in the system. In Fig. 1, the alphabets I_1 through I_{11} represent various item vectors in 3-dimensional user space. Item vectors which are enclosed in the circle are identified as similar items to I_5 by using a similarity measure, whereas the remaining items are dissimilar to I_5. In order overcome the sparsity of I_5, the proposed approach finds the mean of all similar vectors (Given as C in Fig. 1) and add to I_5 so that some of the unrated items of I_5 are filled with ratings. The vector sum of $C + I_5$ is the expanded vector of I_5.

The remainder of the paper is arranged as follows: Sect. 2 presents the work related to CF. Section 3 discusses about the proposed Item Refinement Approach, Sect. 4 discusses about experimental results and Sect. 5 gives conclusions and possible future work.

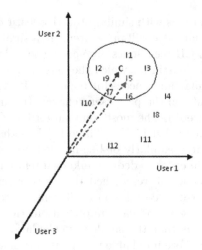

Fig. 1. Geometrical view of item vector expansion technique, $C + I_5$ is the expanded vector of I_5

2 Collaborative Filtering Techniques

Broadly the CF techniques are classified into Neighbourhood based techniques and Matrix factorization based techniques. The brief introduction about the techniques is given below.

2.1 Neighborhood Based CF Techniques

The most common form of CF is the neighborhood-based approach (also known as k Nearest Neighbors). The neighborhood CF techniques can be user based or item based [27]. These kNN techniques identify items that are likely to be rated similarly or like-minded people with similar history of rating or purchasing, in order to identify unknown relationships between users and items. Merits of the neighborhood-based approach are intuitiveness, sparing the need to train and tune many parameters, and the ability to easily explain to the user the reasoning behind a recommendation [1]. In user based CF, given a target user, the recommender system considers the users who share similar rating pattern with the target user as his neighbors and use their ratings in order to predict the unrated items of the target user [17]. Most commonly used metric for calculating similarity between users is Pearson Correlation coefficient and is formulated as given in (1)

$$W_{u,v} = \frac{\sum_{i \in I} (r_{u,i} - \bar{r}_u) \times (r_{v,i} - \bar{r}_v)}{\sqrt{\sum_{i \in I} (r_{u,i} - \bar{r}_u)^2} \times \sqrt{\sum_{i \in I} (r_{v,i} - \bar{r}_v)^2}} \tag{1}$$

where I is the set of items rated by both users u and v. $r_{u,i}$ is the rating provided by user u for item i. $(\bar{r}_u)$ is the average rating of user u. $W_{u,v}$ can be between

-1 and $+1$. Predictions can be made based on the weighted average of known ratings as defined in (2)

$$P_{a,i} = \bar{r}_a + \frac{\sum_{u \in U} W_{(u,a)} \times (r_{u,i} - \bar{r}_u)}{\sum_{u \in U} W_{(u,a)}} \tag{2}$$

where $P_{a,i}$ is the predicted value of active user a and item i and U is the set of top k similar users of the active user a. Whereas in case of item based CF [23], given a target item, items which share similar ratings with the given target item are used for prediction computations. Most commonly used metric for calculating similarity between items is adjusted cosine similarity and is formulated as given in (3)

$$Sim_{i,j} = \frac{\sum_{u \in U(i) \cap U(j)} (r_{u,i} - \bar{r}_u) \times (r_{u,j} - \bar{r}_u)}{\sqrt{\sum_{u \in U(i) \cap U(j)} (r_{u,i} - \bar{r}_u)^2} \times \sqrt{\sum_{u \in U(i) \cap U(j)} (r_{u,j} - \bar{r}_u)^2}} \tag{3}$$

where $U(i)$ is the set of users who rated for item i. $r_{u,i}$ is the rating provided by user u for item i. $(\bar{r}_u)$ is the average rating of user u. $Sim_{i,j}$ can be between -1 and $+1$. Predictions can be made based on the weighted average of known ratings defined in (4)

$$P_{u,i} = \frac{\sum_{j \in S(i)} Sim_{i,j} \times (r_{u,j})}{\sum_{j \in S(i)} Sim_{i,j}} \tag{4}$$

where $P_{u,i}$ is the predicted value of user u and item i and $S(i)$ is the set of top k similar items of item i. Slope One [11] algorithm works on an intuitive principle of a 'popularity differential' between items for users. It calculates how much better one item is liked by another. Given the training data, the deviation between any two items can be calculated by the formula defined in (5)

$$Dev_{i,j} = \sum_{u \in U_{(i,j)}} \frac{r_{u,i} - r_{u,j}}{card(U_{(i,j)})} \tag{5}$$

where $Dev_{i,j}$ is the average rating deviation of items i and j and $U_{i,j}$ is the set of users who rated for both items i and j and $Card(U_{(i,j)})$ is the cardinality of the set S. Predictions can be calculated based the formula given in (6)

$$P_{u,i} = \bar{u} + \frac{1}{Card(R_u)} \times \sum_{j \in R_u} Dev_{(i,j)} \tag{6}$$

where $P_{u,j}$ is the prediction for user u and item i and R_u is the set of items rated by the user u.

Central to most kNN approaches is similarity measure which identifies neighbors of users or items. Even though kNN techniques are popularly used in many commercial recommender systems, they are fully dependent on similarity measures. Similarity measures consider common items/users, may fail to identify neighbours if sufficient ratings are not available in the data set.

2.2   Matrix Factorization Based CF Techniques

Latent factor models are an alternative approach that tries to explain the ratings by characterizing both items and users on, say, 20 to 100 factors inferred from the ratings patterns. Latent factor models are based on Matrix factorization techniques. Examples include pLSA [7], neural networks [28], Latent Dirichlet Allocation [2], or models that are induced by Singular Value Decomposition (SVD) on the user-item ratings matrix [22]. They have gained popularity because of their effectiveness in reducing the size of the rating database. Unlike the neighbor based techniques, the matrix factorization techniques [10] use the existing ratings to learn a model with k latent variables for user and for items. Thus these techniques represent the entire data set in m × k dimensions space where m, n are the number of users, items and $k < n$ is the number of latent features. So the predicted rating of item i for user u can be computed as an inner product of user-factor vector for user u and item factor vector for item i.

2.3   Relevance Feedback

Relevance feedback is a process to modify a search process to improve accuracy of search results by incorporating information obtained from prior relevance judgments [21]. The relevance feedback techniques are classified into three categories. They are Implicit Feedback, Explicit Feedback and Pseudo Relevance Feedback. Rochio Pseudo Relevance Feedback (PRF) [19] via query expansion (QE) is an effective technique for boosting the overall performance in Information Retrieval (IR) domain [14].

Pseudo Relevance Feedback technique is applied in two phases. In the first phase, documents are ranked for a query and top ranked documents are considered to be relevant and in the second phase, the original query vector is expanded by adding potentially related terms from those relevant documents [4].

3   Proposed Item Refinement Approach

For each item, the remaining item vectors are classified as relevant (RE) and non relevant (NR) based on item-item similarity values. Rochio Relevance Feedback (RRF) technique is used to expand item vectors. By expanding an item vector, we bring in the ratings given by users for relevant items of the target item, if it is not rated.

The expanded ratings are of two kinds: rating of an item that already exists in original rating matrix and the rating that is unavailable in the original rating matrix. The deviation of the ratings in the former one is called as Expansion Deviation. The deviations are adjusted in order to refine the expansion of item vectors. The proposed module works in three stages namely Item-Item Similarity Computation Phase, Item Expansion Phase and Item Expansion Refinement Phase. Although CF techniques take different kinds of input, matrix representation is considered here.

Definition 1. User - Item Matrix R

If there are m users who have given ratings for n items, then the ratings data can be represented as an $m \times n$ matrix with rows representing users and columns representing items. The matrix is called user-item rating matrix R. Each element $R_{u,i}$ is an ordinal value which ranges from R_{min} to R_{max}. Unrated values are considered to be zero.

A sample rating matrix R is shown in Table 1.

Table 1. User-item rating matrix R

Items

	I1	I2	I3	I4	I5	I6
U1	1	2	0	5	0	0
U2	0	3	0	0	5	0
U3	0	0	5	0	0	2
U4	0	1	0	0	0	0

3.1 Item-Item Similarity Computation Phase

The proposed approach works in three phases. The first phase of item refinement is to find out the items which are relevant (similar) to the target item. Many similarities namely Pearson correlation coefficient, Cosine Similarity, Adjusted Cosine Similarity etc., are available in the literature. The similarity measure considered here is based on user co-occurrence in item vectors. The reason behind selecting co-occurrence of items is that an item vector with very less number of ratings would be able to get neighbors. Therefore cold start item vectors get expansions. The relationship/similarity between item vector i and item vector j is defined as

$$S_{i,j} = \begin{cases} 1 & if \quad |U_i \cap U_j| > 0 \\ 0 & otherwise \end{cases} \tag{7}$$

where U_i is the set of users who have rated for item i and $|U_i \cap U_j|$ represents cardinality of the set $(U_i \cap U_j)$.

Definition 2. Item-Item Similarity Matrix S

In the given $m \times n$ user item rating matrix R, Item-Item similarity matrix can be represented as an $n \times n$ symmetric matrix S. The matrix rows and columns represent items and each $S_{i,j}$ represents the binary similarity of item i and item j. The Item-Item similarity matrix of R is shown in Table 2.

Table 2. Item-Item similarity matrix S

Items						
	I1	I2	I3	I4	I5	I6
I1	0	1	0	1	0	0
I2	1	0	0	1	1	0
I3	0	0	0	0	0	1
I4	1	1	0	0	0	0
I5	0	1	0	0	0	0
I6	0	0	1	0	0	0

3.2 Item Expansion Phase

In Item-Item Similarity matrix S, for each item i, the remaining items are classified in to RI, the set of relevant items if $S_{i,j} = 1, \forall j \in 1, 2, 3, ...n$ and NRI, the set of non relevant items if $S_{i,j} = 0, \forall j \in 1, 2, 3, ...n$. Relevance Feedback based Rocchio's model with no negative feedback has the following steps,

- For each item vector, classify other item vectors into Relevant (RI) and Non-Relevant (NRI) based on initial relevance decision.
- Each item vector in the feedback set RI is represented as a feedback item vector.
- The representation of the target item vector is refined by taking a linear combination of the initial item vector and the set of feedback item vectors.

Each item vector of the original rating matrix is expanded based on Rochio Relevance Feedback [19] as formulated in Eq. (8)

$$I_{new} = \alpha \times I_{old} + \beta \times \frac{1}{|RI|} \sum_{I_j \in RI} I_j - \gamma \times \frac{1}{|NRI|} \times \sum_{I_k \in NRI} I_k \qquad (8)$$

where I_{old} is any item vector of original rating matrix and I_{new} is corresponding expanded item vector. The constants α, β and γ are the weights assigned to original item vector, set of relevant item vectors and set of non relevant item vectors of I_{old} respectively. RI and NRI are sets of relevant and non-relevant items of I_{old}. The cardinality of RI and NRI are $|RI|$ and $|NRI|$ respectively. Since only the relevant item vectors of the given item vector are considered, β is set as 1 and γ as 0. Since ratings given by users in the original matrix have to be considered, α is set as 1.

Equation given in (8) is interpreted as

$$IEM_{u,i} = R_{u,i} + \frac{1}{|RI|} \times \sum_{j \in RI} R_{u,j} \qquad (9)$$

where RI is the set of feedback items of i and n is the total number of feedback items of item i. If an item has many neighbors, the contribution by those neighbors is high. In order to give high priority to less popular items, we divide the neighbors contribution by the popularity score of the current item. So Eq. (9) is modified as

$$IEM_{u,i} = R_{u,i} + \frac{1}{pop_i} \times \frac{1}{|RI|} \times \sum_{j \in RI} R_{u,j} \tag{10}$$

where pop_i is defined to be

$$pop_i = \frac{|U_i|}{\max_j |U_j|} \tag{11}$$

where U_i is the set of users rated for item i and $|U_i|$ is the cardinality of the set U_i. Popularity of an item i is the ratio of number of users rated the item i to the number of users rated for the maximum rated item.

Definition 3. Item Expansion Matrix, IEM

For the given rating matrix R, the Item Expansion Matrix can be represented as $m \times n$ matrix, IEM. Rows of IEM represent users and the columns represent items. Each $IEM_{u,i}$ represents the expanded rating of user u for item i. Each expanded rating $IEM_{u,i}$ is greater than the corresponding original rating $R_{u,i}$. IEM for the matrix R is shown in Table 3.

Table 3. Item Expansion Matrix IEM

Items	I1	I2	I3	I4	I5	I6
U1	11.500	4.000	0.000	9.500	6.000	0.000
U2	4.500	4.667	0.000	4.500	14.000	0.000
U3	0.000	0.000	11.000	0.000	0.000	17.000
U4	1.500	1.000	0.000	1.500	3.000	0.000

3.3   Item Expansion Refinement Phase

Given the set of actual and expanded ratings pair $<R_{u,i}, IEM_{u,i}>$, the deviation between them is called Item Vector Expansion Deviation which shows how the rating is increased by the ratings assigned to the relevant items of i by the user u. i.e., each is a linear combination of the rating given by user u for item i and the average of ratings of all relevant items of i by the user u. The deviation between them is called Item Vector Expansion Deviation. The deviation refinement is carried out in two steps:

Step 1

Given the set of actual and expanded ratings $<R_{u,i}, IEM_{u,i}>$ the Item Expansion Deviation is computed as the deviation between ratings of R and IEM and is as

$$IEDM_{u,i} = IEM_{u,i} - R_{u,i} \tag{12}$$

Definition 4. Item Expansion Deviation Matrix, IEDM

For the given rating matrix R, the Item Expansion Deviation Matrix can be represented as $m \times n$ matrix ($IEDM$). The matrix rows represent users and the columns represent items. Each $IEDM_{u,i}$ represents rating deviation of user u for item i.

$IEDM$ of the sample rating matrix R is shown in Table 4. According to the proposed technique, the values in each row of $IEDM$ shows how a user is deviating between an item and all its relevant items. For example from Table 4, total deviations of user U_1 is $10.500 + 2 + 4.500 = 17.000$ (for known ratings of U_1 in the original rating matrix R). Average Expansion deviation of U_1 is thus 5.666. In the same way the expanded ratings of the unrated items of U_1 namely I_3, I_5 and I_6 would also contain such deviations. So we subtract the average rating deviation of U_1 from all the three values in order to get the ratings that would be given by U_1 for those unrated items.

Table 4. Item Expansion Deviation Matrix, $IEDM$

Items						
	I1	I2	I3	I4	I5	I6
U1	10.500	2.000	0.000	4.500	6.000	0.000
U2	4.500	1.667	0.000	4.500	9.000	0.000
U3	0.000	0.000	6.000	0.000	0.000	15.00
U4	1.500	0.000	0.000	1.500	3.000	0.000

Definition 5. User Average Deviation Vector, UAD_u

For the given $m \times n$ User-Item rating matrix R, UAD_u is represented as a $1 \times m$ row vector, which is calculated as

$$UAD_u = \sum_{i=1}^{n} \frac{IEDM_{u,i}}{q}, \forall i \in \{1, 2, ...n\}, R_{u,i} > 0 \qquad (13)$$

where q is total number of rated items of user u in the original matrix R.

Step 2

Once the Expansion of item vectors and the user average deviations are calculated, Item Deviation Refinement can be made by subtracting the user average deviation from IEM. Deviation refinement for user u and item i can be computed as

$$IDRM_{u,i} = IEM_{u,i} - UAD_u, \quad \forall R_{u,i} = 0, i \in \{1, 2, ...n\} \qquad (14)$$

where UAD_u is the user average rating deviation of user u.

Definition 6: Item Deviation Refinement Matrix, IDRM

Given the set of actual and expanded ratings $<R_{u,i}, IEDM_{u,i}>$, Item Deviation Refinement Matrix $IDRM$ can be filled with refined values in case of unknown ratings of R. $IDRM$ will retain all known ratings of R. i.e., the ratings given by users in the original matrix are retained in $IDRM$. Each $IDRM_{u,i}$ represents the refined rating of user u for the item i if it is not rated in R.

The $IDRM$ matrix for the matrix R is shown in Table 5.

Table 5. Item Deviation Refinement Matrix ($IDRM$)

Items	I1	I2	I3	I4	I5	I6
U1	1.000	2.000	−5.667	5.000	0.3333	−5.667
U2	−8.337	3.000	−5.333	−.8333	5.000	−5.333
U3	−10.500	−10.500	5.000	−10.500	−10.500	2.000
U4	1.500	1.000	0.000	1.500	3.000	0.000

The proposed technique takes two phases namely offline phase and online phase. In the offline phase, the expansion refinement is made and Item Deviation Refinement matrix ($IDRM$) is generated. In the online phase, based on the matrix created, predictions are made.

4   Experimental Results

Details about the data sets used, accuracy of predictions and computational complexities are described here. In collaborative recommender systems, it is crucial to accurately predict how much a user likes an item if less number of ratings is available. In addition, processing time required to make rating prediction is also an important issue.

4.1   Data Set Used

The experiments are conducted on standard data set namely Movielens (www. Movielens.com). Two data sets from Movielens namely Movielens100k and Movielens1M are considered for experiments. The data set contains ratings given for movies in the range 1 to 5. Movielens100k consists of ratings given by 943 users for 1684 items. The total number of ratings available is 1,00,000. Movielens1M contains ratings given by 6000 users for 4000 movies. The total number of ratings available is 1 million ratings. To scale down Movielens1M data set, a set of 1000 users is randomly selected with 2200 common items. Five fold cross validation is done to validate the proposed approach.

In order to prove the efficiency of Item Refinement technique, Item vectors with very less ratings are considered. In order to discuss about the efficiency of item expansion approach, two data sets from the given data sets are created. From the given data sets, item vectors with maximum of 3 ratings, items with a maximum of 5 ratings and items with maximum of 8 ratings are considered for comparison of results.

4.2   Evaluation Metrics Used

According to Herlocker et al. [6], metrics evaluating recommendation systems can be broadly classified into two broad categories: predictive accuracy metrics, such as Mean Absolute Error (MAE), and its variations; classification accuracy metrics, such as precision, recall and F1-measure.

In this work, three predictive accuracy metrics as discussed in [29] are used. The first one is the hit-rate (HR) which is the ratio of the number of hits to the size of the test set. The predicted rating is referred as a *hit* if its rounded value is equal to the actual value in the test set. A HR value of 1.0 implies that all the ratings are predicted correctly. One limitation of the hit-rate measure is that it is indifferent to the distance to actual rating in case of a miss. This limitation is addressed by the Mean Absolute Error, MAE, which penalizes each miss by the distance to actual rating. The last method is Root Mean Square Error, RMSE, a measure that emphasizes large errors compared to MAE measure. The algorithm with low MAE an RMSE values is considered to be the best algorithm. The metrics are defined as

$$HR = \frac{no.\ of\ hits}{n} \tag{15}$$

$$MAE = \sum_{i=1}^{n} \frac{|P_i - A_i|}{n} \tag{16}$$

$$RMSE = \sqrt{\sum_{i=1}^{n} \frac{(P_i - A_i)^2}{n}} \tag{17}$$

where P_i is the predicted rating, A_i is the actual rating, n is the total number of ratings.

Since the motive of the proposed approach is to improve the quantity of recommendations, the metric 'coverage' which is the ratio of number of predictions made to the total number of known ratings is used. The formula for coverage is as follows,

$$coverage = \frac{N_{r,s}}{N_r} \tag{18}$$

where $N_{r,s}$ is the number of ratings predicted, N_r is the number of known ratings in the test set. The algorithm with high HR and coverage is considered to be the best algorithm.

4.3   Comparison of Algorithms

This section shows that when item vectors are expanded the number of ratings in it increases in a meaningful way which in turn improves quality of predictions. In order to prove the efficiency of item vector expansion, kNN approach for item based collaborative filtering detailed in [23] is considered. kNN algorithm is applied on base line data set as well as expanded item vectors. The procedures are called as kNN_Item, kNN_IE respectively. Further the proposed approach is compared with kNN based user collaborative filtering [17], SVD based collaborative filtering [22] and collaborative error reflected models, ERR [9].

Comparisons of results for various metrics in case of item expansion of Movielens100k are shown in Table 6. From Table 6, it is understood that kNN_IE seems to produce better prediction accuracy with respect to MAE, RMSE, HIT and Coverage. The improvement in the quality of predictions is due to expansion of item vectors. The values given in bold letters show the improvement of prediction accuracy produced by the proposed vector expansion technique.

Table 6. Comparison of results of item expansion in Movielens100k

	No. of ratings per item ≤ 3			
	MAE	RMSE	HIT	COVERAGE
kNN_Item	1.1310	1.5054	0.0879	0.1985
kNN_IE	**1.0260**	**1.3456**	**0.1522**	**0.3551**
kNN_User	1.1260	1.4456	0.1122	0.2351
SVD	1.1160	1.5456	0.1322	0.2451
ERR	1.0460	1.3956	0.1422	0.2851
	No. of ratings per item ≤ 5			
	MAE	RMSE	HIT	COVERAGE
kNN_Item	1.1045	1.2054	0.0879	0.2085
kNN_IE	**1.0045**	**1.1156**	**0.1922**	**0.4251**
kNN_User	1.1002	1.2164	0.1217	0.2185
SVD	1.0156	1.1978	0.1349	0.2725
ERR	1.0136	1.1278	0.1649	0.2925
	No. of ratings per item ≤ 8			
	MAE	RMSE	HIT	COVERAGE
kNN_Item	1.0985	1.1986	0.1079	0.2385
kNN_IE	**1.0002**	**1.1226**	**0.2222**	**0.4891**
kNN_User	1.1085	1.2036	0.1179	0.2585
SVD	1.0245	1.2046	0.1479	0.2456
ERR	1.01415	1.191046	0.1479	0.3286

Comparisons of results for various metrics on Movielens1M are shown in Table 7.

Table 7. Comparison of results in item expansion in Movielens1M

	No. of ratings per item ≤ 3			
	MAE	RMSE	HIT	COVERAGE
kNN_Item	1.1510	1.5063	0.0759	0.1856
kNN_IE	**1.1030**	**1.3446**	**0.1642**	**0.2811**
kNN_User	1.1540	1.4663	0.0984	0.1756
SVD	1.1210	1.4263	0.1259	0.2456
ERR	1.1110	1.3663	0.1559	0.2656
	No. of ratings per item ≤ 5			
	MAE	RMSE	HIT	COVERAGE
kNN_Item	1.1055	1.3424	0.1079	0.1985
kNN_IE	**0.9934**	**1.1356**	**0.2022**	**0.3921**
kNN_User	.9989	1.2387	0.1479	0.2235
SVD	1.1055	1.3424	0.1579	0.2985
ERR	1.0015	1.2824	0.1959	0.2355
	No. of ratings per item ≤ 8			
	MAE	RMSE	HIT	COVERAGE
kNN_Item	1.0025	1.2316	0.1129	0.2645
kNN_IE	**0.9876**	**1.1226**	**0.2422**	**0.4291**
kNN_user	1.0012	1.2134	0.1676	0.2545
SVD	1.0123	1.3316	0.1729	0.2945
ERR	0.9912	1.1316	0.2229	0.3645

From Tables 6 and 7 it can be observed that the proposed item expansion improves quality of predictions on different data sets.

4.4 Comparing Computational Complexities

This section, discusses about the computational complexity of the proposed method. High computational complexity is often needed to enhance the predictions. In the model based view, the computational complexity can be split into complexity in offline phase and complexity in online phase [5]. The offline phase includes computation of $IDRM$, and similarity computations. $IDRM$ can be pre-computed since it does not change quickly. Periodically updating this matrix would be more than enough. Online phase includes prediction computations of n test items. In online phase, computational complexity needed to predict ratings for n items using k nearest neighbours is $O(kn)$.

5   Conclusions and Future Work

The goal of this work is to improve the performance of collaborative Filtering. In many real recommender systems, large portion of item vectors and user profiles have less number of ratings. Moreover similarity computations largely depend on ratings available in data set and leads to inferior predictions if sufficient data is not available. Therefore including additional ratings to rating data set is a key of success to improve quality of predictions. Item Refinement approach is presented in the paper includes additional ratings with an objective of improving prediction quality. The results show that the proposed technique outperforms the benchmark collaborative filtering technique. Further, most of the computations can be done in advance as an offline component and thus on-line computational complexity is same as that of kNN technique. The proposed approach does not work if the existing item vector is not having even a single rating. The reason is that an item vector with no rating does not take any similar item and so the proposed method fails. Instead of binary similarity, different similarity measures can be used to improve accuracy of predictions still better.

References

1. Bell, R.M., Koren, Y.: Improved neighborhood-based collaborative filtering. In: KDD Cup and Workshop at the 13th ACM SIGKDD International Conference on Knowledge Discovery and Data Mining (2007)
2. Blei, D.M., Ng, A.Y., Jordan, M.I.: Latent Dirichlet allocation. J. Mach. Learn. Res. **3**, 993–1022 (2003)
3. Breese, J.S., Heckerman, D., Kadie, C.: Empirical analysis of predictive algorithms for collaborative filtering. In: Proceedings of the Fourteenth Conference on Uncertainty in Artificial Intelligence, pp. 43–52. Morgan Kaufmann Publishers Inc. (1998)
4. Cao, G., Nie, J.Y., Gao, J., Robertson, S.: Selecting good expansion terms for pseudo-relevance feedback. In: Proceedings of the 31st Annual International ACM SIGIR Conference on Research and Development in Information Retrieval, pp. 243–250. ACM (2008)
5. Deshpande, M., Karypis, G.: Item-based top-n recommendation algorithms. ACM Trans. Inf. Syst. (TOIS) **22**(1), 143–177 (2004)
6. Herlocker, J.L., Konstan, J.A., Terveen, L.G., Riedl, J.T.: Evaluating collaborative filtering recommender systems. ACM Trans. Inf. Syst. (TOIS) **22**(1), 5–53 (2004)
7. Hofmann, T.: Latent semantic models for collaborative filtering. ACM Trans. Inf. Syst. (TOIS) **22**(1), 89–115 (2004)
8. Kantor, P.B., Rokach, L., Ricci, F., Shapira, B.: Recommender Systems Handbook. Springer, Heidelberg (2011)
9. Kim, H.N., El-Saddik, A., Jo, G.S.: Collaborative error-reflected models for cold-start recommender systems. Decis. Support Syst. **51**(3), 519–531 (2011)
10. Koren, Y., Bell, R., Volinsky, C.: Matrix factorization techniques for recommender systems. Computer **42**(8), 30–37 (2009)
11. Lemire, D., Maclachlan, A.: Slope one predictors for online rating-based collaborative filtering. In: SDM, vol. 5, pp. 1–5. SIAM (2005)

12. Lika, B., Kolomvatsos, K., Hadjiefthymiades, S.: Facing the cold start problem in recommender systems. Expert Syst. Appl. **41**(4), 2065–2073 (2014)
13. Lops, P., de Gemmis, M., Semeraro, G.: Content-based recommender systems: state of the art and trends. In: Ricci, F., Rokach, L., Shapira, B., Kantor, P.B. (eds.) Recommender Systems Handbook, pp. 73–105. Springer, Heidelberg (2011)
14. Miao, J., Huang, J.X., Ye, Z.: Proximity-based Rocchio's model for pseudo relevance. In: Proceedings of the 35th International ACM SIGIR Conference on Research and Development in Information Retrieval, pp. 535–544. ACM (2012)
15. Pazzani, M.J., Billsus, D.: Content-based recommendation systems. In: Brusilovsky, P., Kobsa, A., Nejdl, W. (eds.) The Adaptive Web. LNCS, vol. 4321, pp. 325–341. Springer, Heidelberg (2007). doi:10.1007/978-3-540-72079-9_10
16. Rashid, A.M., Karypis, G., Riedl, J.: Learning preferences of new users in recommender systems: an information theoretic approach. ACM SIGKDD Explor. Newsl. **10**(2), 90–100 (2008)
17. Resnick, P., Iacovou, N., Suchak, M., Bergstrom, P., Riedl, J.: GroupLens: an open architecture for collaborative filtering of netnews. In: Proceedings of the 1994 ACM Conference on Computer Supported Cooperative Work, pp. 175–186. ACM (1994)
18. Resnick, P., Varian, H.R.: Recommender systems. Commun. ACM **40**(3), 56–58 (1997)
19. Rocchio, J.J.: Relevance feedback in information retrieval (1971)
20. Ronen, R., Koenigstein, N., Ziklik, E., Nice, N.: Selecting content-based features for collaborative filtering recommenders. In: Proceedings of the 7th ACM Conference on Recommender Systems, pp. 407–410. ACM (2013)
21. Salton, G.: The Smart Retrieval System—Experiments in Automatic Document Processing. Prentice-Hall, Upper Saddle River (1971)
22. Sarwar, B., Karypis, G., Konstan, J., Riedl, J.: Application of dimensionality reduction in recommender system-a case study. Technical report, DTIC Document (2000)
23. Sarwar, B., Karypis, G., Konstan, J., Riedl, J.: Item-based collaborative filtering recommendation algorithms. In: Proceedings of the 10th International Conference on World Wide Web, pp. 285–295. ACM (2001)
24. Schafer, J.B., Konstan, J., Riedl, J.: Recommender systems in e-commerce. In: Proceedings of the 1st ACM Conference on Electronic Commerce, pp. 158–166. ACM (1999)
25. Shani, G., Brafman, R.I., Heckerman, D.: An MDP-based recommender system. In: Proceedings of the Eighteenth Conference on Uncertainty in Artificial Intelligence, pp. 453–460. Morgan Kaufmann Publishers Inc. (2002)
26. Son, L.H.: Dealing with the new user cold-start problem in recommender systems: a comparative review. Inf. Syst. **58**, 87–104 (2016)
27. Su, X., Khoshgoftaar, T.M.: A survey of collaborative filtering techniques. Adv. Artif. Intell. **2009**, 4 (2009)
28. Tang, H., Shim, V.A., Tan, K.C., Chia, J.Y.: Restricted Boltzmann machine based algorithm for multi-objective optimization. In: 2010 IEEE Congress on Evolutionary Computation (CEC), pp. 1–8. IEEE (2010)
29. Yildirim, H., Krishnamoorthy, M.S.: A random walk method for alleviating the sparsity problem in collaborative filtering. In: Proceedings of the 2008 ACM Conference on Recommender Systems, pp. 131–138. ACM (2008)

Computational Intelligence

Interaction Model of Service Discovery Using Visa Processing Algorithm and Constrained Application Protocol (CoAP)

S. Umamaheswari and K. Vanitha$^{(\boxtimes)}$

Dr. G. R. Damodaran College of Science, Coimbatore, India
{umamaheswari.s, vanitha.k}@grd.edu.in

Abstract. Service Discovery is a process of identifying the compatible resources of a transaction to result it in a comfort zone. The environment of Internet of Things would be a mixture of hardware and software independent components. Offering a service in an IoT architecture is challenging problem based on participant registration and logs. A model for Integrating Physical world devices in constrained web environments using Constrained Application Protocol (CoAP) together with an end-to-end IP and RESTful Web Services based architecture is proposed. We suggest a Visa Processing Algorithm for an IoT environment to maintain time constraint participants and for anonymity finding. The proposed model is an integrating Constrained Application Protocol with the algorithm for service discovery issues particularly handling anonymous entity.

Keywords: Service discovery · Internet of Things · CoAP · Visa processing · Anonymity

1 Introduction

The scope of the Internet is broadened beyond its limits (e.g. servers and routers) and rapidly growing borders (e.g. Personal Digital Assistants, personal computers and smart phones) to trillions of tiny embedded devices (e.g. Sensors) in the physical world. This enormous growth brings more possibilities and challenges to the Internet, such as seamless integration of constrained devices with the Web. The term Internet of Things (IoT) was coined more than ten years before by the researchers but has emerged into the majority public view only more recently. There is an assumption that the IoT will completely change the usage of computer networks for the next 10 to 100 years. Internet of Things represents a common model for the capability of network devices to sense and collect data from the world around us, and then share that data over the Internet where it can be manipulated and exploited for various interesting purposes.

The existing web technologies and standards can be reused at the maximum level to merge the cyber-world and physical world. The network of physical objects that include embedded technology to communicate and sense or with their internal states or the external environment is termed as the Internet of Things (IoT) and this is emerging as the Web of Things (WoT) [2]. The devices become IP enabled and connected to the Internet and enabled to communicate a language to provide interoperability. Re-designing and

© Springer Nature Singapore Pte Ltd. 2016
S. Subramanian et al. (Eds.): CSI 2016, CCIS 679, pp. 99–106, 2016.
DOI: 10.1007/978-981-10-3274-5_8

optimizations in application protocols is required to implement machine-to-machine (M2M) applications over constrained environments on the IoT. Even though HTTP is widely used with Web Services, it is not the only protocol for M2M communication. The Internet Engineering Task Force (IETF) published a RESTful web transfer protocol called Constrained Application Protocol (CoAP) [3]. CoAP comprises of several HTTP functionalities which can be suited for the constrained environment M2M applications on the IoT, meaning it takes into account the less processing power and energy restrictions of small embedded devices, such as sensors.

Figure 1 illustrates the various applications of Internet of Things (IoT). IoT assumes that the underlying network devices and the associated technology can operate semi-intelligently and many times automatically. Maintaining the mobile devices connected to the Internet can be hard enough to make them smarter. People have ample number of needs that seek an IoT system to accommodate or be configurable for different situations and preferences. Hence, with all those challenges overcome, if people become too dependent on this concept and the technology is not highly robust, any technical knowledge in the system can cause serious system and/or financial damage.

Fig. 1. Few applications of IoT

Section 3 describes the Visa Processing algorithm. Section 4 presents the service discovery process in IoT environment. Section 5 briefs about the implementation of the Visa Processing algorithm using CoAP and the Sect. 6 gives the concluding remarks.

2 Literature Review

Internet Engineering Task Force (IETF) has Constrained Restful environments (CORE) working group published a restful web transfer protocol named as constrained application protocol (COAP) in June 2010 [3]. COAP offers different http functionalities which have been designed again for mobile to mobile applications over constrained environments on the IoT that takes into an account low processing power and energy restrictions of small embedded devices, such as sensors. COAP also offers a number of features that HTTP not covers, such as IP multicast support, built-in resource discovery, native push model, and asynchronous message exchange [13, 14]. The passenger can automatically get a service during his travel using a FlyPS model and discussed the service discovery issues in Airport Scenario [13]. A device/person enters into a network is treated as anonymous user rather a malicious user. Registrations and logs are maintained in a Pervasive environment [14].

3   Visa Processing

A. VISA Processing Techniques for an IoT Constrained Environment. VISA is a special type of document provided to an individual for the entry/exit to a region for being some amount of time. Visa provider is usually being the authority body of the Region of Interest (ROI). The Person who receives Visa can be a visitor/guest to the region. The task of issuing the visa for a new arrival device or a user is taken care by an IoT Server by referring the Guest Registry. Visa document is categorized into multitude ways in many nations in real time context. The IoT Constrained Environment is assumed as a region of interest to be visited and the device/user requires getting a visa from the Server. Visitor visa referred as Short-stay visa is the essential one. The Server identifies a new arrival to the boundary region is touched. On-arrival visa is issued with a time stamp.

(a) Entry Visa Processing Algorithm
Algorithm 1: Entry-Visa Processor
(Entry permit and verification in an IoT Space)
Begin

    Let (variables)

    IR → IoT Region

    B(i) → set of boundaries for an IR

    X → an anonymous device

    IS → IoT Server

    GR → Guest Log Registry

    If (X is in IS) then

      {

          declare 'X' as a registered device

      }

    Else (X is in GR)

      {

          declare 'X' as a guest device

          determine a finite time 'T'

      }

End

(b)   Exit Processing Algorithm for an IoT Space

Algorithm 2: Exit-Visa Processor

(exit processing from an IoT Space)

Begin

Let (variables are assumed from Algorithm 1)

If (X expires) then

{

deactivate the services provided to 'X'

mark in 'X' as removed in GR

}

Else (X volunteers to quit)

{

cross-verify the GR for 'X'

mark in 'X as removed in GR

}

End

The algorithm verifies every user in order to permit into the Pervasive space. An interface installed with Entry processor verifies the registered users. Similarly the exit processor verifies to permit a device/resource to leave the network. The entries of GR are used for this purpose. Consequently, the nodes pass the signals in the network. The signals are repeated generally in order to keep the information safe. This is referred as replicated signals. As a refinement and filtering process of aggregation the replicated signals are to be identified.

4   Service Discovery in IoT Environments

Service discovery is an growingly an important issue as we move towards realizing pervasive systems. How does the user find and access to a particular service from the plethora of services that may potentially be available around him at any moment? After discovered a service, a next important issue is how the user interacts with them. An IoT environment is a mixture of devices, people and services. A life span of a service completes only when the result of the communication is proper. Usually a service layer in the form of middleware resolves this issue.

(i)   **Constrained Environment**

A Constrained Environment is a workspace with low powered hardware devices, with minimum networking capabilities. The existing web services are extended with REStful architecture to provide a service to the IoT devices. The use of web services (web APIs) on the Internet has become pervasive in most applications and depends on the fundamental Representational State Transfer [REST] architecture of the Web.

(ii) **CoAP**

According to IETF Standards Release, The Constrained Application Protocol (CoAP) is a specific web transfer protocol for use with restricted nodes and the restricted (e.g., low-power, lossy) networks. The nodes frequently have 8-bit microcontrollers with small amounts of primary memory, while constrained networks such as IPv6 over Low-Power Wireless Personal Area Networks (6LoW-PANs) frequently will have high-level packet error rates and a usual throughput of 10 s of Kbit/s. The protocol is generated for machine-to-machine (M2M) applications such as smart home and building automation. CoAP provides a request/response interaction model between domain endpoints, supports discovery of services in built-in mode and resources, and includes key concepts of the Internet such as URIs and Web media types. CoAP is designed to easily interface with HTTP for integration of the Web when addressing specialized requirements such as multicast support, very less overhead, and simplicity for constrained environments

CoAP is an Application Layer service discovery protocol for constrained environments. Resource discovery in built-in, IP support for multicast routing, native push model, and asynchronous message exchange are other special features of CoAP that are not offered by HTTP (Fig. 2).

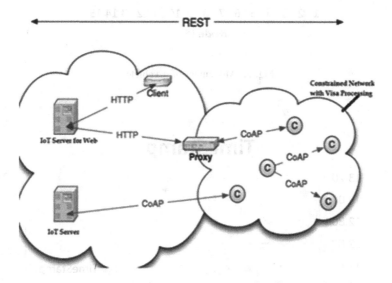

Fig. 2. An IoT environment using CoAP interactions

5   Implementation

The Model is realized in a Simulated Environment. The Contiki Operating System for CoAP environment is used to implement the Visa Processing Algorithm. An Agricultural logistics is the application scenario. The datasets are used from Great IoT,

Sensor and other Data Sets Repositories for Japanese Logistics Sensor Data. Among the various attributes observed in the original dataset we consider the boundaries in the form of longitudes and the validity of the endpoint. As the Visa Processing Algorithm concentrates on the boundary crossing and verification, expiry timestamp of the endpoint, we observe the nodes for the above.

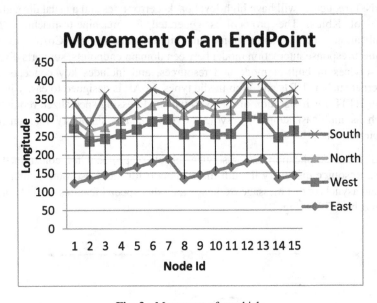

Fig. 3. Movement of a vehicle

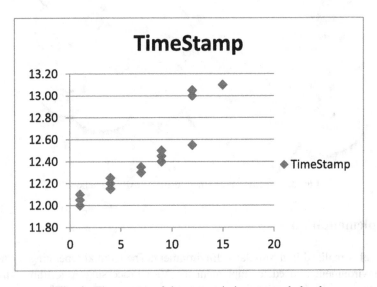

Fig. 4. Time stamp of data transmission at a node level

The Fig. 4 illustrates the timestamps of a single endpoint in a virtually fenced farm. As the time increases linearly, we observe the node with a synchrounous data transmission.

The CoAP environment is created in a simulated scenario with 15 nodes of IoT devices. The node specification and configuration is assumed as random and self-configured. The heterogeneous devices are deployed to interact each other and to the base station. No cluster/cluster head is fixed. The Web of Things is assumed as an extension of HTTP in the form of CoAP. The following graph in Fig. 3 illustrates the movement of an Endpoint in various locations of the farm. The Endpoint is assumed as a moving vehicle to serve the farming practices. The locations are the landmarks deployed with specific static sensor nodes. When the vehicle enters into the node coverage, the interaction is initiated and accounted. The four directions are measured for virtual fencing environment as constrained. The interaction itself is a service discovery process of the environment.

6 Conclusion

The proposed model and algorithm recommend the integration of CoAP in IoT environment. The authors highlighted the use of CoAP in pervasive agriculture environment. CoAP is used on the sensor nodes in the vehicle container to retrieve resources in both directions. This model is an IP based solution to integrate sensor networks used in the agriculture farm, highlighting the use of the CoAP protocol for the retrieval of sensor data during the traversal of a vehicle. The proposed model defines service discovery for IoT at the middleware level. The entry and exit visa processing algorithms provides a virtual fencing to a pervasive space. The key research issue is to handle unidentified and unauthorized entities into the pervasive environments. The first time enquiry or missing registrations are to be focused for future avenues. The real time implementation of a precision farming with IoT devices interacting to each other using our model is also planned.

References

1. Shelby, Z.: Embedded web services. J. IEEE Wireless Commun. **17**(6), 52–57 (2010)
2. Zeng, D., Guo, S., Cheng, Z.: The web of things: a survey (invited paper). J. Commun. **6**(6), 424–438 (2011)
3. Shelby, Z., Hartke, K., Bormann, C., Frank, B.: Constrained application protocol (CoAP). IETF Internet-Draft 08, November 2011. http://tools.ietf.org/html/draft-ietf-core-coap-08
4. Dritsas, S., Gritzalis D., Lambrinoudakis, C.: Protecting privacy and anonymity in pervasive computing: trends and perspectives. 23(3) (2006)
5. Shelby, Z.: Constrained application protocol (CoAP). Internet Eng. Task Force CoRE Working Group, Technical report, February 2011. http://tools.ietf.org/html/draft-ietf-core-coap-07

6. Aijaz, A., Hamid Aghvami, A.: Cognitive machine-to-machine communications for internet-of-things: a protocol stack perspective. IEEE Internet Things J. **2**(2), 103–112 (2015)
7. An Introduction to Internet of Things. http://www.cisco.com/web/solutions/trends/iot/introduction_to_IoT_november.pdf
8. Jara, A.J., Ladid, L., Skarmeta, A.: The internet of everything through IPv6: an analysis of challenges, solutions and opportunities. J. Wirel. Mob. Netw. Ubiquit. Comput. Dependable Appl. **4**(3), 97–118 (2013)
9. Palattella, M., et al.: Standardized protocol stack for the internet of (important) things. IEEE Commun. Surv. Tutorials **15**(3), 1389–1406 (2013)
10. Miao W., Ting L., Fei L., Ling S., Hui D.: Research on the architecture of internet of things. In: IEEE International Conference on Advanced Computer Theory and Engineering (ICACTE), Sichuan Province, China, pp. 484–487 (2010)
11. Guthikonda, R.T., Chitta, S.S., Tekawade, S., Attavar, T.: Comparative Analysis of IoT Architectures, April 2014
12. Liang, Z., Chao, H.-C.: Multimedia traffic security architecture for the internet of things. IEEE Netw. **25**(3), 35–40 (2011)
13. Vanitha, K., Sasikala, R.: Pervasive computing in airport scenario: service discovery issues. In: First International Conference in South Asia on Global Manufacturing Systems and Management (ICGMSM-2011), Coimbatore Institute of Technology and AIMMRG and BIRC, August 2011
14. Radhamani, G., Vanitha, K., Amphawan, A.: Anonymity handling and sensor object modeling for pervasive environments using VISA processing. In: International Conference on Future Communications and Computing ICFCCS' 2014 (ISI, Scopus Indexed, IGI-Global), Dubai, May 2014

A Constraint Based Heuristic for Vehicle Routing Problem with Time Windows

G. Poonthalir[1(✉)], R. Nadarajan[1], and S. Geetha[2]

[1] Department of Applied Mathematics and Computational Sciences,
PSG College of Technology, Coimbatore 641 004, India
poonthalirk@gmail.com, nadarajan_psg@yahoo.co.in
[2] Department of Computer Applications,
Government Arts College, Udumalpet, India
geet_shan@yahoo.com

Abstract. Vehicle Routing Problem with Time Windows (VRPTW) is a well known combinatorial optimization problem. Many solution strategies are proposed to solve VRPTW. In this work, a constraint based (COB) heuristic approach is proposed to solve VRPTW. The algorithm uses two stages to solve. In the first stage, priority based decomposition is introduced that partitions the customers based on spatial constraints with a devised priority metric. In the second stage, an urgency based orientation is introduced based on temporal constraints to direct the customers along the route. The routes obtained are improved for optimality. The proposed algorithm is tested on Solomon's benchmark data sets and implemented using *MATLAB 7.0.1*. The results obtained are found to be competitive with other well known methods.

Keywords: Vehicle Routing Problem with Time Windows · Combinatorial optimization · Heuristics · Spatial constraint · Temporal constraint

1 Introduction

The distribution of goods or products from a set of production centers to a set of customers is of great economic and social importance. The focal point of distribution logistics is to minimize the overall cost of delivery. However, cost minimization alone is not a fundamental strategy. There should be a good tradeoff between minimizing cost and maximizing other competitive performance metrics such as quality, delivery, service, etc. Also the success of a distribution depends on planning of routes and prompt delivery of goods. These problems are extensively studied under Vehicle Routing Problems (VRP) and are used widely in the areas of Transportation Engineering, Logistics and Operations Research. VRP is a NP hard Combinatorial Optimization Problem (COP). It is used to design optimal routes for serving customers who are all geographically distributed with some demand. They are to be served by vehicles with homogenous capacity stationed at depot. The objective of VRP is to optimize the route cost and to serve all customers. In VRPTW, a set of vehicles with homogenous capacity is to be routed from a central depot to serve customers with some demand who are geographically dispersed. Each customer has a time window with two quantities,

© Springer Nature Singapore Pte Ltd. 2016
S. Subramanian et al. (Eds.): CSI 2016, CCIS 679, pp. 107–118, 2016.
DOI: 10.1007/978-981-10-3274-5_9

early time and late time within which the vehicle should serve them. The objective is to reduce the route cost without violating the time constraints.

This paper aims in introducing a solution technique using the spatial and temporal constraint of customers. The algorithm aims to solve the problem in two stages. In the first stage, the problem is decomposed into smaller sub problems using the spatial location with a devised priority metric. This reduces the complexity of the problem. Then, each sub problem is solved separately and integrated to get the solution to the original problem using a devised algorithm urgency based orientation.

2 Literature Survey

The solution strategies used to solve VRPTW are varied and can be classified into solutions obtained using exact algorithms, Heuristics and Meta heuristics. Exact algorithms are used to get accurate results, but often take exponential time. [1] proposed an exact algorithm and devised a method based on shortest path problem with resource constraints. Their algorithm required a tight deadline constraint. [2] suggested a set of heuristics and improvement methods for VRPTW that has its origin in solving VRP. Similar to Clarke and Wright's savings heuristic, Solomon described a heuristic that works in parallel route construction along with time window constraints. Insertion heuristic and nearest neighborhood heuristic described by Solomon were widely used by researchers. These form the base for new heuristic to emulate. [3] described a procedure to solve the delivery schedules for a food processing company with greedy look ahead heuristic with less number of vehicles. [4] proposed a reactive tabu search with adjustable tabu list during the iteration process. A simulated annealing approach combined with a local search can be found in [5]. Work on genetic algorithms can be found in [6] where they formulated VRPTW as a set partitioning problem with genetic algorithm. [7] proposed a hybrid Meta heuristic which uses the Ant Colony Optimization (ACO) and the tabu search. ACO is used to arrive at a solution and tabu is used to explore new solutions, a local improvement method is also used. Particle Swarm Optimization (PSO) is also used to get solution for VRPTW. [8] extended their work of VRP with PSO to solve VRPTW with PSO.

3 Problem Description

VRPTW can be formulated on undirected graph $G = (V, E)$ with a set of vertices V representing the set of N customers $\{C_1, C_2, \ldots, C_N\}$ where $C_i \in V \backslash \{0\}$ are distributed in the Euclidean plane with co-ordinates (x_i, y_i) Let E be the set of edges connecting the vertices. Each edge is associated with a cost $\cos t_{ij}$ connecting the customers C_i and C_j. Each customer i has a demand d_i. The depot V_0 has a fictitious demand zero and has a fleet of M vehicles stationed at the depot with homogenous capacity Q. Each customer has a service time p_i and time window (a_i, b_i) where a_i is the early time, b_i is the late time. The vehicle should start from the depot, service the customer with the required demand exactly once within the time window and return to the depot without violating the capacity constraint of the vehicle. The service time p_i is the time taken for servicing the customer. If a vehicle arrives before early time a_i there is a wait time w_i given by $w_i = t_i - a_i$ where t_i is the total time taken to reach the customer i. Each vehicle should arrive before the late time b_i. TT_k is the total allowable time taken by vehicle k, $k \in M$. A decision variable x_{ijk} is introduced to know the customers i and j who are served by a particular vehicle k

$$x_{ijk} = \begin{cases} 1 \text{ if } k \text{ travels from i to j} \\ 0 \text{ otherwise} \end{cases}.$$

The mathematical formulation of the problem as given by (Yu et al. [7]) is

$$\min \sum_{i=0}^{N} \sum_{j=0}^{N} \sum_{i \neq jk=1}^{M} \cos t_{ij} x_{ijk} \tag{1}$$

$$\sum_{k=1}^{M} \sum_{i=1}^{N} x_{0jk} \leq M \tag{2}$$

$$\sum_{j=1}^{N} x_{ijk} = \sum_{j=1}^{N} x_{jik} \leq 1 \text{ for } i = 0, \ k = 1, 2, \ldots, M \tag{3}$$

$$\sum_{j=1}^{N} x_{j0k} = 1 \quad \text{for } k = 1, 2, \ldots M \tag{4}$$

$$\sum_{j=1}^{N} x_{0jk} = 1 \text{ for } k = 1, 2, \ldots M \tag{5}$$

$$\sum_{i=1}^{N} x_{ihk} - \sum_{j=1, j \neq i}^{N} x_{hjk} = 0 \text{ for } h = 1, 2, \ldots N \, k = 1, 2, \ldots M \tag{6}$$

$$\sum_{i=1}^{N} \sum_{j=0, j \neq i}^{N} x_{ijk} \leq Q \text{ for } k = 1, 2, \ldots M \tag{7}$$

$$\sum_{i=1}^{N} \sum_{j=0, j \neq i}^{N} x_{ijk}(t_{ij} + p_i + w_i) \leq TT_k \text{ for } k = 1, 2, \ldots, M \tag{8}$$

$$\sum_{k=1}^{M} \sum_{i=0, i \neq j}^{N} x_{ijk} = (t_{ij} + p_i + w_i + t_i)x_{ijk} \leq t_j \text{ for } j = 1, 2, \ldots, N \tag{9}$$

$$a_i \leq t_i \leq b_i \text{ for } 1 = 1, 2, \ldots, N \tag{10}$$

Time when a vehicle arrives at customer j is calculated as,

$$t_j = \sum_{i=0}^{N} \sum_{k=1}^{V} x_{ijk}(t_{ij} + p_i + w_i + t_i) \text{ for } j = 1, 2, \ldots, N \qquad (11)$$

Equation (1) is the objective function to minimize the total travel length under several constraints. Constraint (2) sets the maximum number of vehicles that start from the depot. Constraint (3) is the travel constraint, (4) and (5) shows that the vehicle starts and ends at the depot, constraint (6) shows that the vehicle enters to serve customer h and leaves customer h is the same, constraint (7) describes the capacity of the vehicle, constraint (8) is the maximum travel time. (9) and (10) are the time window constraints. Constraint (11) shows the time taken to reach a customer.

4   Proposed Work

The proposed work aims to solve VRPTW in two stages. The first stage is the decomposition stage where the problem is decomposed and each individual sub problem is solved separately. With decomposition, the dimensionality of the problem to be solved is reduced. This decomposition can be related to decide the number of vehicles used to service the customer. Based on the geographical location and demand of the customer a priority metric is devised to carry out the decomposition. Though the dimensionality of the problem is reduced it is still difficult to solve, hence a Constraint Based heuristic is introduced that exploit the temporal constraints of the customer to find the route. The resulting route is feasible and a local search is employed to improve the route for optimality.

4.1   Decomposition Based on Priority

The process by which a complex problem is divided into parts that are easier to conceive and solve is referred as decomposition. It aims in breaking down a problem to minimize the static dependencies among the parts. The methodology is as follows.

Construction of Groups: The customers $C_1, C_2, \ldots, C_n$ are distributed in the Euclidean plane with co-ordinates (x_i, y_i) *for* $i = 1, 2, \ldots, N$. These N customers are decomposed into K groups. The minimum value of K is taken as one so at least one group is formed. The total number of groups formed is $K \leq M$. Using Eq. (12) the number of groups K is determined.

$$K = \begin{cases} \left\lceil \dfrac{\sum_{i=1}^{N} d_i}{Q} \right\rceil, & \sum_{i=1}^{N} d_i > Q \\ 1, & \text{otherwise} \end{cases} \qquad (12)$$

The number of vehicles based at a depot is M. After decomposition, each group will have n_j number of customers assigned to it, satisfying the condition $\sum_{j=1}^{K} n_j = N$.

Selection of Seeds: The number of groups to be formed corresponds to number of vehicles required to serve the customers. An initial set of seeds, S_k *for* $k = 1, 2, \ldots, K$ are determined. The distance $dist_i$ *for* $i = 1, 2, \ldots, N$ is calculated for N customers from the depot. These are arranged in non decreasing order as $dist_1 > dist_2 \ldots > dist_N$. Roughly above half the number of seeds $\lceil \frac{S_K}{2} \rceil$ are chosen from the farthest set of customers and the remaining $(S_K - \lceil \frac{S_K}{2} \rceil)$ set are chosen from the nearest set of customers from the depot. These seeds play an important role in forming the groups that spans the entire Euclidean space.

Formation of Groups: With the seeds got, an initial grouping of the customers is made. The customers $C_1, C_2, \ldots, C_n$ are assigned to corresponding seed $S_1, S_2, \ldots, S_k$ based on a priority metric. A priority metric is devised based on the customers demand and distance. The proposed work decomposes the customer based on the demand; this will pack the customers tight within the group, which aims in the minimal use of vehicles. Two attributes of customers namely the demand and the distance are considered for calculating the priority metric. It assigns the customer with high demand and low distance to have higher priority, so it is included in the group first. These customers are given preference because if the higher demand customer with less distance is grouped first then it is easy to pack the customers with less demand which indirectly aid in the reduction of vehicle. A priority metric P_{ij} for N customers to K seeds $S_1, S_2, \ldots, S_k$ are calculated using Eq. (13).

$$P_{ij} = \frac{dist_{ij}}{demand_i} \quad for \, i = 1, 2, \ldots, N, \; j = 1, 2, \ldots, K \tag{13}$$

$$dist_{ij} = \sqrt{(x_i - x_j)^2 + (y_i - y_j)^2} \tag{14}$$

This grouping will pack the customer tightly. In each group, the number of customers is limited by the capacity of the vehicle. For a group L_i, $i = 1, 2, \ldots, K$, $L_i \subseteq V$ and $\#L_i = n_j$ such that $\sum_{i=1}^{K} \#L_i = N$. Then each group should satisfy the Eq. (15)

$$\sum_{r=1}^{n_i} d_r \leq Q \quad \forall i = 1, 2, \ldots, K \tag{15}$$

Let $(x_1, y_1), (x_2, y_2), \ldots, (x_{n_i}, y_{n_i})$ be the co-ordinates of the customer in a set $L_i \subseteq V$, $i = 1, 2, \ldots, K$. The seeds are updated as in Eq. (16).

$$Z_i = \sum_{i=1}^{n_i} x_i / n_i \text{ and } Z_j = \sum_{i=1}^{n_i} y_i / n_i \tag{16}$$

Where $n_i = \#L_i$, $i \in K$.

After seeds are updated, the groups are regenerated to obtain the optimal grouping of customers. After the grouping, based on the time constraint the customers are checked for service by a corresponding vehicle. If any customer left without service, then the number of seeds is arbitrarily incremented. The process of grouping the customers is iterated until convergence, till all the customers are allocated to the group and there is no change in the customers allocated to the groups. This decomposes the graph G into disjoint sub graphs $G_1, G_2, \ldots, G_K$.

4.2 Orientation Based on Urgency Metric

Each sub graph G_i, $i = 1, 2, \ldots, K$ has the set of customers to be served. This paper proposes urgency metric to direct the customers along the route. Since customers are already grouped together based on spatial constraint, the task is to serve the customers by the vehicle within the specified time window of the customer. Urgency metric is devised based on the late time, wait time and the distance.

Earliest Late Time Rule (ELTR): Within each group, the customers are initially arranged in ascending order of late time.

Wait Time Rule (WTR): For any customer C_w, the wait time w_w is calculated as $w_w = a_w - t_l$ where a_w is the early time, t_l is the travel time till C_w and $a_w > t_l$. The customer with the least/no wait time is selected, assuming to be an urgent customer for servicing.

Closest Distance Rule (CDR): When more than one customer is having the least wait time, then those customers are pipe lined by CDR to identify the closest one, considering that customer as the most urgent.

Methodology. Each sub graph G_i, $i = 1, 2, \ldots, K$ obtained through decomposition is used for routing. For each group, $L_i, i \in K$, the customers in each group C_r, $r \in n_i$ where $i = 1, 2, \ldots, K$ are arranged in non decreasing order of their late time b_r, where $b_1 < b_2 \ldots < b_{n_i}$. Let $H_i, i \in K$ be a list constructed within each group with customers arranged with the above ordering of late time. A typical list has customers arranged as $C_1, C_2, \ldots, C_l, C_w, C_r, \ldots, C_{n_i}$, $n_i = \#L_i$. The list H_i is used as a base list to be consulted to validate whether all customers are visited. The customers with identical time window and cannot be served in a group are transferred to the next nearest group and checked for constraint satisfaction. Since hard time window is considered, to serve a customer C_i, the total travel time t_i while serving C_i should be $a_i \leq t_i \leq b_i$. The selection of customers for servicing is based on two factors, wait time and distance. The wait time of all customers in a list H_i is calculated and stored in list H_i'. The customer with least wait time and smallest distance is selected for service. Equation (17) will decide the servicing of customer C_w by a vehicle after C_l

$$visit(C_w) = \begin{cases} 1, & w_w \leq 0 \text{ and } b_w - t_l < p_w \\ 0, & w_w > 0 \end{cases} \tag{17}$$

If $visit(C_w)$ is 0 then service to customer C_w can be delayed. For any customer C_r, a list PC_r having the set of all probable customers (PC_{rj}), $r \in n_i$, $r \neq j$, where $0 < j < n_i$ is formed. To decide the customer to be served after customer C_l the customers with wait time less than or equal to w_w of C_w are selected as probable customers and stored in PC_l. If there is more number of customers having the least wait time, then CDR is followed to resolve the situation. Let $\{C_1^{\#}, C_2^{\#}, \ldots, C_M^{\#}\}$ be the list of customer in PC_l with least wait time than C_w. The customer C^* is selected using CDR from $C_m^{\#} \forall m = 1, 2, \ldots, M$ using Euclidean distance from C_l to $C_m^{\#} \forall m = 1, 2, \ldots, M$.

4.3 Improvement Procedure

The route formed at the end of the Constraint Based heuristic will be a feasible route. When constructing the route using urgency metric the concentration is made to serve all customers within the time windows. So, the wait time of the customer plays a major role in deciding the service of customer along the route. The proposed method builds the route with the criteria that instead of spending the time in waiting for servicing a customer, the vehicle can serve any other customer with no wait time. This is the basic principle for devising the urgency metric. This in turn guides to serve all customers. Since the objective is to reduce the total travel time the route obtained is tried for optimality with a compromise in wait time. So an improvement method is used to optimize the route. A simple greedy heuristic is used to improve the route. A customer is selected and inserted in the cheapest position. If the cost is less than the calculated cost, route is updated.

VRPTW introduces many complications in improving the routes as hard time window constraints are involved. The precedence in terms of the time window constraints should be taken into account. One way to get an infeasible route is when a violation in precedence is found in terms of time window constraints. To eliminate this, a preprocessing step is included to eliminate the introduction of infeasible route by the construction of probable list. Since the probable list of all customers to be served is maintained for each customer, alternate route can be formulated without considering all possibilities which reduces the time complexity. But, if there is a change in the path, then again based on the time window alternate route are to be constructed.

4.4 Theoretical Analysis of Proposed Algorithm

Lemma 1: Route obtained using constraint based heuristic is a feasible route.

Proof: Let L_i, $\forall i \in K$ be the groups formed. Let the initial ordering formed within each cluster based on the late time in L_i be $H_i \forall i \in K$. To prove that the route obtained is feasible, it is enough to prove that the solution satisfies the constraints Eqs. (2)–(10). To satisfy the constraints related to travel and vehicles used, the clustering L_i, $i \in K$ is analogous to the number of vehicles (K) used, where $K \leq M$. This implies that each cluster L_i, $i \in K$ is served by one vehicle. If C_r, $r = 1, 2, \ldots, n_i \forall i \in K$ is the set of all customers in each L_i, then all customers within L_i are served by the same vehicle which starts and ends at the depot. To satisfy the constraint related to capacity of a vehicle, the initial clustering L_i is formed as specified in Eq. (7) hence it is satisfied.

To verify whether the temporal constraints are satisfied, the ordering of customer C_r within each L_i is done with the ELTR. Each customer C_r, r = 1, 2, ..., n_i $\forall i \in K$, has a time window $[a_r, b_r]$. If a typical customer ordering has customer C_l followed by customer C_w then $b_l < b_w$ which fixes a priority on the order of the customer to be served based on their late times. Hence the routes formed are feasible.

Lemma 2: Customers within each group are tightly packed using priority metric.

Proof: Let $C_1, C_2, ..., C_N$ be the customers with demand $d_1, d_2, ..., d_N$. If the customers are put in a group using Euclidean distance alone as a measure, then the customers with minimum distance from the group center will be considered for including in the group. This may be true for those customers with lesser demand also. Then the customer with less demand also gets packed in a group that leads to

$$\sum_{r=1}^{n_i} d_r + \beta = Q \; \forall i = 1, 2, ..., K$$ where n_1 is a group. Let β be the remaining space

within a group. Since demands of the customers are not considered there is a chance of customer with greater demand (d_g) not packed in to the group and not able to find place in any groups because of higher demand i.e. $d_g > \beta$ of each vehicle. In this case there may be a chance of a few customers with greater demand not be able to fit in a vehicle that leads to an additional vehicle to be allocated.

If the customers are arranged in decreasing order of their demand and increasing order of distance, then the high demanding customers are packed first leaving space for less   demanding   customers   to   be   accommodated   at   the   end.

$$\sum_{r=1}^{n_i} d_r + \beta = Q \; \forall i = 1, 2, ..., K$$ Where $\beta \approx 0$. Since these customers' demands are less

they can be accommodated in any of the groups, hence the need for another vehicle is eliminated. Thus the customers within each group are tightly packed.

5   Results and Discussions

The proposed work is tested with the Solomon's 56 test cases. They are C1, R1, RC1, C2, R2 and RC2. The customers C1 and C2 are clustered customers. RC1 and RC2 are both remote and clustered customers, R1 and R2 are remotely placed customers. Problems C1, R1 and RC1 have a narrow time window, so less number of customers is served in this category when compared to C2, R2 and RC2. The algorithm is coded with MATLAB 7.0.1 with Intel core 2 quad processor. The proposed method's efficiency is tested by comparing the overall cost and cumulative total number of vehicles with the published results [8]. Table 1 project that the proposed method works better. To supplement the strength of the algorithm it is also compared against Meta heuristics like tabu search, genetic algorithm and Ant colony optimization algorithms. Table 2 shows the comparison obtained.

Table 2 describes the cumulative total number of vehicles used and the cumulative total distance travelled by the vehicle. The average Relative Percentage Deviation

Table 1. Comparison of the proposed method with other heuristic

Data set		[9]	This work
C1	NV[a]	10	10
	TD[b]	**828.38**	**828.6**
C2	NV	3	3
	TD	596.63	590.92
R1	NV	12.42	**12**
	TD	1233.34	1225.37
R2	NV	3.09	**4.5**
	TD	990.99	990.08
RC1	NV	12	**11.63**
	TD	1403.74	1411.68
RC2	NV	3.38	**5.6**
	TD	1220.99	1169.04
Cumulative Total Vehicle	TNV	420	**445**
Cumulative Total Distance	TD	58927	**58272**

[a]N - Number of Vehicles; [b]TD - Total Distance.

Table 2. Comparison of the results of the proposed method with other methods

Data set		[10]	[11]	[4]	[12]	[13]	[14]	[7]	This work
C1	N[a]	10	10	10	10	10	10	10	10
	TD[b]	838	828.45	828.38	828.94	828.38	832.13	841.92	828.6
C2	N	3	3	3	3	3	3	3.3	3
	TD	589.90	590.30	591.42	589.93	589.86	589.86	612.75	590.52
R1	N	12.60	12.25	12.17	12.50	12	12.16	13.1	12
	TD	1296.83	1216.70	1204.19	1268.42	1217.73	1211.55	1213.16	1225.37
R2	N	3.00	3.00	2.73	3.09	2.73	3	4.6	4.5
	TD	1117.70	995.38	986.32	1055.90	967.75	1001.12	952.3	990.08
RC1	N	12.10	11.88	11.88	12.25	11.63	12.25	12.7	11.63
	TD	1446.20	1367.51	1397.44	1396.07	1382.42	1418.77	1415.62	1411.68
RC2	N	3.40	3.38	3.25	3.38	3.25	3.37	5.6	5.6
	TD	1360.60	1165.62	1229.54	1308.31	1129.19	1170.93	1120.37	1169.04
All	N	422	416	411	423	408	418	470	445
	TD	62572	57993	58502	60651	59597	58476	57799	58272

[a]N - Number of Vehicles; [b]TD - Total Distance.

(RPD) is used to compare the relative performance of the algorithms with the proposed method. RPD is calculated using Eq. 18. Table 3 projects the RPD's obtained for the proposed method with other methods.

Table 3. RPD's obtained by other method with the proposed method

Other methods	C1	C2	R1	R2	RC1	RC2
[10]	−1.12	0.105	−5.510	−11.418	−2.386	−14.079
[11]	0.018	0.037	0.7125	−0.532	3.229	0.293
[4]	0.026	−0.15	1.758	0.381	1.019	−4.920
[12]	−0.041	0.100	−3.393	−6.233	1.118	−10.645
[13]	0.026	0.111	0.627	2.307	2.116	3.529
[14]	−0.424	0.111	1.140	−1.102	−0.499	−0.162
[7]	−1.582	−3.627	1.006	3.967	−0.278	4.344

Table 4. Effects of Improvement heuristic on the proposed method

Dataset	ACBI	ACAI	RPI
C1	836.3126	828.6	0.92221
C2	615.78	590.5163	4.102723
RC1	1482.738	1411.675	4.792655
RC2	1234.263	1169.039	5.284431
R1	1291.758	1225.369	5.139442
R2	11844.7	10890.95	8.052125

$$RPD = \frac{(obtained\_result - Known\_result)}{Known\_result} \times 100 \qquad (18)$$

For some data sets, the proposed method works better compared to the published results. As seen in the results, the deviations if exists are lesser in case of clustered data sets when compared to remote and clustered data sets. Since the algorithm proceeds

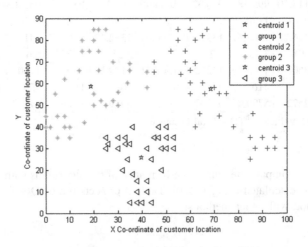

Fig. 1. Grouping obtained for data set C201

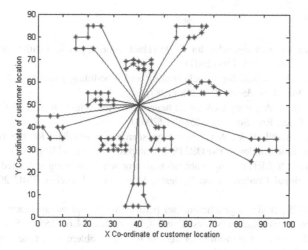

Fig. 2. Routing for C101 data set

with grouping the customers with the spatial constraint the customers are tightly packed within a group, and not much deviation is obtained for clustered set of customers.

Table 4 presents the solutions obtained by the proposed method before and after improvement. ACBI is the Average Cost obtained before the improvement heuristic, ACAI is the Average Cost obtained after the improvement heuristic where *RPI* is given as, $RPI = \frac{(ACBI-ACAI)}{ACBI} \times 100$.

This work use the priority metric that groups the customers based on both demand and distance. Figure 1 shows the grouping obtained at the end of decomposition based on priority metric for the data set C201 and routing for C101 is shown in Fig. 2. Since the customers are tightly packed, as the vehicle number increases, more grouping is formed which increases the cost of the solution. After the construction of the feasible route, the route is improved for optimality.

6 Conclusion

In this paper a constraint based heuristic method is proposed for solving the VRPTW. The method is used to solve the VRPTW in two stages. In the first stage the problem is decomposed into sub problems and each individual sub problem is solved separately. A priority metric is used to group the customers based on the spatial constraint. The second stage uses an urgency metric based on a set of rules that orients the customers along the route. The results obtained using the proposed work is tested on the data sets of Solomon. To assess the efficiency of the algorithm the obtained results are compared with the other well known heuristics. The total distance and the total number of vehicles used are also tabulated.

References

1. Azi, N., et al.: An exact algorithm for single vehicle routing problem with time window. Eur. J. Oper. Res. **178**(3), 755–766 (2007)
2. Solomon, M.: Algorithms for vehicle routing and scheduling problem with time window constraints. Oper. Res. **35**(2), 254–265 (1987)
3. Ioannou, G., et al.: A greedy look-ahead heuristic for the vehicle routing problem with time windows. J. Oper. Res. Soc. **52**, 523–537 (2001)
4. Chiang, W., Russell, R.A.: A reactive tabu search metaheuristic for the vehicle routing problem with time windows. INFORMS J. Comput. **9**(4), 417–430 (1997)
5. Lin, S.W., et al.: Vehicle routing problems with time windows using simulated annealing. In: IEEE International Conference on Systems, Man and Cybernetics SMC 2006, pp. 8–11 (2006)
6. Alvarenga, G.B., et al.: A genetic and set partitioning two-phase approach for the vehicle routing problem with time windows. Comput. Oper. Res. **34**(6), 1561–1584 (2007)
7. Yu, B., et al.: A hybrid algorithm for vehicle routing problem with time windows. Expert Syst. Appl. **38**, 435–441 (2011)
8. Ai, T.J., Kachitvichyanukul, V.: A particle swarm optimization for vehicle routing problem with time windows. Int. J. Oper. Res. **6**(4), 519–537 (2009)
9. Caseau, Y., Laburthe, S.: Heuristics for large constrained vehicle routing problems. J. Heurist. **5**, 281–303 (1999)
10. Potvin, J.Y., Bengio, S.: The vehicle routing problem with time windows part II: genetic search. INFORMS J. Comput. **8**(2), 165–172 (1996)
11. Taillard, É., et al.: A tabu search heuristic for the vehicle routing problem with soft time windows. Transp. Sci. **31**(2), 170–186 (1997)
12. Schulze, J., Fahle, T.: A parallel algorithm for the vehicle routing problem with time window constraints. Ann. Oper. Res. **86**, 585–607 (1999)
13. Lau, H.C., et al.: Vehicle routing problem with time windows and a limited number of vehicles. Eur. J. Oper. Res. **148**(3), 559–569 (2003)
14. Tan, K.C., et al.: A hybrid multi-objective evolutionary algorithm for solving vehicle routing problem with time windows. Comput. Optim. Appl. **34**(1), 115–151 (2006)

Soft Clustering Based Missing Value Imputation

P.S. Raja[✉] and K. Thangavel

Department of Computer Science, Periyar University, Salem 636 011, India
psraja5@gmail.com, drktvelu@yahoo.com

Abstract. Preprocessing is one of the steps in Data Mining, which involves Noise removal, Identification of outlier, Normalization, Data transformation, Handling missing values, etc. Missing value is a common problem in large datasets. Most frequently used method to handle missing values by statistical is discarding the instances with missing values. Sometime deletion of instances with missing values cause loss of essential information, which affects the performance of statistical and machine learning algorithms. This paper focuses on handling missing values using unsupervised learning techniques. Rough K-Means based missing value imputation was proposed and compared with K-Means, Fuzzy C-Means based imputation methods. The experimental analysis is carried out on two data sets Lung Cancer and Cleveland Heart data sets. The proposed method achieves the best accuracy for some of the datasets.

Keywords: K-Means · Fuzzy C-Means · Rough K-Means · Missing value imputation

1 Introduction

Missing data refer the incompleteness in attribute values in a data set which may occur for different reasons. The reasons for missingness are considered as the failure of equipment, measurement error, ignorance and data corrupted. The observation with missing values are difficult to analysis in the statistical modeling [1].

Decision making is one of the process of Data Mining which refers to extracting the knowledge from large amounts of data. It depends on the rich complete data, from which accurate information can be extracted. But many data sets available for research contain missing data [2]. Some of the preprocessing methods are data transformation, data normalization and missing values handling. One of the most significant issues of missing values in a large data sets are imputing the missing values and improve the accuracy. Imputing the missing values is really a challenging task in data mining preprocessing [2].

Based on the pattern and the mechanism of the missingness, the characteristics of the missingness are identified. In the data set, the way of missing occurred is identified by the pattern and the mechanism shows that the probabilistic definition of the missing values [3]. The missing data mechanisms are Missing at Random (MAR), Missing Completely at Random (MCAR) and Missing Not at Random (MNAR).

© Springer Nature Singapore Pte Ltd. 2016
S. Subramanian et al. (Eds.): CSI 2016, CCIS 679, pp. 119–133, 2016.
DOI: 10.1007/978-981-10-3274-5_10

Some of the methods for handling missing values are as follows [4], (i) Ignore the tuple is usually done when the class label is missing. (ii) Fill in the missing value manually is another approach, but it is time-consuming and may not be feasible for a given large data set with many missing values. (iii) One may use the most probable value to fill all the missing value [5].

In the past few years, clustering based missing values imputation was popularized among the researcher. Basically clustering techniques are unsupervised method use to group similar objects in the same cluster and based on the information on the cluster the missing values are imputed [6]. Sometime the data points that are similar distance to two or more clusters are simply placed exactly any one of the clusters. It leads to being the wrong conclusion of the missing values.

Simplicity and speed are the main advantages of K-Means clustering algorithm in large dataset [7]. In K-Means clustering algorithm group similar objects in the same cluster and dissimilar objects in different clusters. The main drawback of the K-Means clustering have not yielded the same result on each run, it depends on the initial random assignments of the K centroid. It affects the experimental results because each run it imputes different value. To obtain the optimal result, we need some efficient method to impute the missing values [8].

In Fuzzy C-Means the data point which belong to more than one cluster are represented by membership function whose values are between zero and one [9–11]. In Fuzzy C-Means the data point which belong to the particular cluster is represented by the high degree of belongingness. In Rough K-Means, it enables the data point they belong to one or more clusters, but the degree of membership value is not available. The Proposed Rough K-Means Centroid based imputation methods is compared with K-Means and Fuzzy C-Means based imputation methods.

The rest of the paper is organized as follows. Section 2 discusses clustering algorithms related to missing values imputation. Section 3 discusses three imputation algorithms based on centroid method and three imputation algorithms based on parameters. Section 4 discusses the experimental analysis. Section 5 discusses the conclusion.

2   Clustering Algorithm Used for Imputation

2.1   K-Means Algorithm

The K-Means Clustering algorithm was developed by MacQueen in 1967. The brief description of K-Means algorithm was given by Hartigan in 1975. The main aim of the K-Means algorithm is partitioning X point into K clusters on the basis of a sample. Clustering is the process of grouping similar data point within the same cluster and dissimilar data point in different cluster.

The K-Means algorithm for clustering X input data points into K cluster, where $X = \{x_i | i = 1, 2, \ldots, n\}$ is the data points, $C = \{C_i | i = 1, 2, \ldots, K\}$, is the cluster, where $0 < K < n$. The dissimilarity measure is used to cluster the data is given as follows:

$$E = \sum_{i=1}^{K} \sum_{x \in C_i} dist(x, m_i)^2 \qquad (1)$$

where x represents the data point in the cluster C_i and m_i is the mean of cluster C_i. The most popular clustering algorithm is K-Means clustering algorithm, which initially start from random assignment of centroid and assign the data item to the cluster by the closest value of centroids. The same process is repeated until there is no more convergence. The details and steps involved in K-Means are detailed in [6].

2.2 Fuzzy C-Means (FCM) Algorithm

In Fuzzy C-Means each data point is characterized by membership, which belongs to all the cluster by a membership function with ranges between zero and one as against K-Means algorithm where the membership function takes either zero or one. Each data point memberships close to unity signify a high degree of similarity between the data point and a cluster while memberships close to zero imply little similarity between the data point and that cluster. In Fuzzy C-Means 'fuzziness' m is the important parameter. One can find the steps involved in Fuzzy C-Means algorithm in [12].

2.3 Rough K-Means Algorithm

Rough set properties were introduced by Zdzislaw Pawlak in the early 1980's. It is a mathematical approach to treat imprecision, uncertain and vagueness information [11]. Based on the properties of rough set and the data mining concepts of classic K-Means clustering approach, Lingras et al. proposed a Rough Cluster algorithm. Peter's Refinements on rough clustering algorithms is used in this work. In rough clustering each cluster contains the lower and upper approximation. Lower approximation is the subset of the upper approximation. The importance of the rough clustering is an object belongs to the lower approximation is one and only membership to that cluster and also that object present in upper approximation. An object belongs to the upper approximation that is not contained in the lower approximation of the cluster is at least the present in more than two upper approximations of the clusters and the actual membership of the data object are missing. In rough clustering Lingras et al. are not considering all the properties of the rough set, but the use following property for rough cluster algorithm [10].

- A data object belongs can belongs to one lower approximation at most.
- A data object is a member of lower approximation, it's also a member of at least one upper approximations.
- A data object that does not belong to any lower approximation is a member of at least two upper approximations.

3 Imputation Algorithms

3.1   Centroid Based Imputation Methods

In centroid based imputation methods, missing values imputation is achieved by the following ways. The original dataset is partitioned into two groups, they are objects without missing attribute values or Complete Data (CData) and objects with missing attribute values or Missing Data (MData). Clustering algorithm is applied to CData to partitioning the dataset into K cluster, which gives $M = \{m_i | i = 1, 2, \ldots, K\}$ centroid is the average value of column vectors of each cluster C_i. The object with missing attribute values in MData is fetched one by one, the missing column value of an object $s_{i,j}$ is replaced by corresponding column values of $m_{i,j}$ centroid where $S = \{s_1, s_2, \ldots, s_n\}$. Then the distance is calculated for the object s_i and the m_i centroid. In the same way K distance is calculated for all possible m centroid and object s_i. Find the minimum distance d_k from K distance, then permanently placed missing column value of an object $s_{i,j}$ by the corresponding column value of $d_{k,j}$ centroid value and placed the object s_i to the d_k cluster [6]. The same process is repeated for all objects in MData.

K-Means Centroid Based Imputation Algorithm.

Algorithm : K-Means Centroid based Imputation Algorithm
Input   : Dataset X of n objects with missing attribute values and number of clusters K, $(K<n)$
Output   : Objects without missing values (Imputed Dataset)

Step 0: Given dataset of n objects $X = \{x_1, x_2, \ldots, x_n\}$, CData $\subset$ X, MData $\subset$ X, Where n is the number of objects, CData contain objects without missing data and MData contains objects with missing attribute values.

Step 1: Initialize the number of cluster K.

Step 2: K-Means Clustering algorithm is applied on CData to partitioning the objects. The distance measure used for K-Means clustering algorithm is follow.

$$E = \sum_{i=1}^{k} \sum_{x \in C_i} |x - m_i|^2$$

Step 3: After the clustering process, replace missing s_j column value with m_j column value and find the minimum distance d between s_i and centroid. Where MData=$\{s_i \mid i=1, \ldots, T\}$, T is the number objects in MData.

$$\vec{d_i} = \min_{k=1,\ldots,K} d(\vec{s_i}, \vec{m_k})$$

Step 4: Replace the s_j missing attribute values with $d_{i,j}$ attribute values and place in the lth cluster.

Step 5: Repeat step 3 and step 4 for all objects in MData.

Step 6: Stop.

In K-Means Centroid based imputation method the non-reference attribute (objects with missing values) values are imputed based on the knowledge of the same cluster. Objects within the cluster are more similar to each other that show the mean values (cluster centroid) of objects in the cluster and the object within the cluster are approximately holds the similar values. Imputing the non-reference attributes from the knowledge of cluster centroid value gives the best optimal values for the missing attribute. Where m is the centroid value here $K = \{1, 2, 3, \ldots, K\}$ Stands for the number of clusters.

$$m_k = \frac{\sum_{x_i \in C_k} x_i}{|C_k|} \tag{2}$$

Equation 2 shows the centroid calculation for K–Means algorithm, it generated the mean value for each cluster. The non-reference attribute in an object is imputed by the corresponding centroid value of that cluster. The accuracy of the imputed value is changed based on the data set and the imputing algorithm. If the variation of the data set is very less, then the objects are very close to each other and that improves the imputation accuracy of the non-reference attribute in centroid based imputation.

Fuzzy C-Means Centroid Based Imputation Algorithm.

Algorithm : Fuzzy C-Means Centroid based Imputation Algorithm
Input : Dataset X of n objects with missing attribute values, membership matrix and number of clusters K, $(K<n)$
Output : Objects without missing values (Imputed Dataset)

Step 0: Given dataset of n objects $X = \{x_1, x_2, \ldots, x_n\}$, CData $\subset$ X, MData $\subset$ X, Where n is the number of objects, CData contain objects with complete data and MData contains objects with incomplete attribute values.

Step 1: Initialize the number of cluster K.

Step 2: Fuzzy C-Means Clustering algorithm is applied on CData. The distance measure for Fuzzy C-Means clustering algorithm is follow.

$$m_k = \frac{\sum_{i=1}^{N} u_{ij}^v \cdot x_i}{\sum_{i=1}^{N} u_{ij}^v}$$

Step 3: After the clustering process, replace missing s_j column value with m_j column value and find the minimum distance d between s_i and centroid. Where MData=$\{s_i \mid i=1,\ldots, T\}$, T is the number objects in MData.

$$\vec{d_l} = \min_{k=1,\ldots,K} d(\vec{s_i}, \vec{m_k})$$

Step 4: Replace the s_j missing attribute values with d_{lj} attribute values and place in the lth cluster.

Step 5: Repeat step 3 and step 4 for all object in MData.

Step 6: Stop.

In Fuzzy C-Means centroid based imputation algorithms the missing attribute values are imputed by the information of the same cluster. The cluster centroid for Fuzzy C-Means is computed by the following equation.

$$m_j = \frac{\sum_{i=1}^{N} u_{ij}^v \cdot x_i}{\sum_{i=1}^{N} u_{ij}^v} \tag{3}$$

where x_i is the object and the u_{ij} represents the membership values for the data object. In Fuzzy C-Means the clusters are formed by the membership function based on the belongingness of the cluster. In Fuzzy C-Means the well separation of data into one cluster is achieved by the membership value of data. In this case, each data contains membership value for all cluster, the highest membership value to the cluster are grouped into one cluster. Then the centroid value for each cluster is computed by using the Eq. 3 and based on the information of the centroid value the missing attribute values are imputed. Based on the closeness of the data in each cluster the algorithm gives the optimal value for the non-reference attribute. In centroid based imputation method the accuracy of the imputed value is increased when there are less variations between the data objects within the cluster.

Rough K-Means Centroid Based Imputation Algorithm.

Algorithm : Rough K-Means Centroid based Imputation Algorithm
Input   : Dataset X of n objects with missing attribute values, threshold, w_l, w_u and number of
             clusters K, $(K<n)$
Output   : Objects without missing values (Imputed Dataset)

Step 0: Given dataset of n objects $X = \{x_1, x_2, ..., x_n\}$, CData $\subset X$, MData $\subset X$, Where n is the number of objects. CData object with complete data and MData object with Incomplete data.

Step 1: Initialize the number of cluster K.

Step 2: Rough K-Means Clustering algorithm is applied on CData. The distance measure for Rough K-Means clustering algorithm is follow.

$$\vec{m}_k = w_l \sum_{\vec{X}_n \in \underline{C_k}} \frac{\vec{X}_n}{|\underline{C_k}|} + w_u \sum_{\vec{X}_n \in C_k} \frac{\vec{X}_n}{|\overline{C_k}|}$$

Step 3: After the clustering process, replace missing s_j column value with m_j column value and find the minimum distance d between s_i and centroid. Where MData=$\{s_i \mid i=1,..., T\}$, T is the number objects in MData.

$$\vec{d}_i = \min_{k=1,...,K} d(\vec{s}_i, \vec{m}_k)$$

Step 4: Replace the s_j missing attribute values with $d_{i,j}$ attribute values and place in the lth cluster.

Step 5: Repeat step 3 and step 4 for all object in MData.

Step 6: Stop.

In Rough K-Means Centroid based imputation algorithms, the missing attribute values are imputed based on the knowledge of the centroid values. The Eq. 4 is the centroid calculation equation for Rough K-Means algorithm. Here m_k holds the centroid value.

$$\vec{m}_k = w_l \sum_{\vec{X}_n \in \underline{C_k}} \frac{\vec{X}_n}{|\underline{C_k}|} + w_u \sum_{\vec{X}_n \in C_k} \frac{\vec{X}_n}{|\overline{C_k}|} \tag{4}$$

$w_l + w_u = 1$ where w_l and w_u represents the importance of lower and upper approximations. Based on the Rough K-Means imputation algorithm, accuracy of the imputing value is improved when the data objects are in less variance. The importance of lower approximation is taken into 0.7 percentages and the importance of upper approximation is taken into 0.3 percentages [10, 11, 15–17].

3.2 Parameter Based Imputation Methods

In parameter based imputation methods, missing values of an object is imputed based on the information about the object in the cluster and some property of that cluster. In Fuzzy C-Means product of membership value with centroid value are used. In Rough K-Means lower bound and upper bound object information are used instead of the cluster centroids [13, 14].

K-Means Parameter Based Imputation Algorithm.

Algorithm : K-Means Parameter based Imputation Algorithm [13].
Input : Dataset X of n objects with missing attribute values and number of clusters K, $(K<n)$
Output : Objects without missing values (Imputed Dataset)

Step 0: Given dataset of n objects $X = \{x_1, x_2, ..., x_N\}$, CData $\subset$ X, MData $\subset$ X, Where n is the number of objects, CData contain objects with complete data and MData contains objects with missing attribute values. Where MData=$\{s_i \mid$ i=1, ..., T$\}$, T is the number of objects in MData.

Step 1: Initialize the number of cluster K.

Step 2: K-Means Clustering algorithm is applied on CData to partitioning the objects. The distance measure used for K-Means clustering algorithm is follow.

$$E = \sum_{i=1}^{k} \sum_{x \in C_i} |x - m_i|^2$$

Step 3: After the clustering process, replace missing s_j column value with m_j column value and find the minimum distance d between s_i and centroid.

$$\vec{d_l} = \min_{k=1,...,K} d(\vec{s_i}, \vec{m_k})$$

Step 4: Place the s_j object in the lth cluster.

Step 5: Apply Nearest Neighbor algorithm to fill in all the non-reference attributes.

Step 6: Repeat step 3 and step 5 for all objects in MData.

Step 7: Stop.

In K-Means Parameter based imputation algorithms the non-reference attributes or the missing attribute of an object are imputed based on the information on the closest object within the cluster. In this method Nearest Neighbor algorithm is used to find the closest object within the cluster, the Eq. 5 shows the distance measure for Nearest Neighbor. Very close to the non-reference object within the cluster is identified by Nearest Neighbor Algorithm and imputed the missing values.

$$dist(X_1, X_2) = \sqrt{\sum_{i=1}^{n} (x_{1i} - x_{2i})^2} \qquad (5)$$

Searching nearest object from the entire data set is complicated process and also increases the computation time. So clustering algorithm is applied to the data set to group the similar object in a cluster, then the Nearest Neighbor algorithm is applied to the cluster to find closest object.

Fuzzy C-Means Parameter Based Imputation Algorithm.

Algorithm : Fuzzy C-Means Parameter based Imputation Algorithm [13].
Input   : Dataset X of n objects with missing attribute values, membership matrix and number of clusters K, $(K<n)$
Output   : Objects without missing values (Imputed Dataset)

Step 0: Given dataset of n objects $X = \{x_1, x_2, ..., x_n\}$, CData $\subset$ X, MData $\subset$ X, Where n is the number of objects, CData contain objects with complete data and MData contains objects with missing attribute values. Where MData=$\{s_i \mid i=1, ..., T\}$, T is the number of objects in MData.

Step 1: Initialize the number of cluster K.

Step 2: Fuzzy C-Means Clustering algorithm is applied on CData. Centroid measure for Fuzzy C-Means clustering algorithm is follows.

$$m_j = \frac{\sum_{i=1}^{N} u_{ij}^v \cdot x_i}{\sum_{i=1}^{N} u_{ij}^v}$$

Step 3: After the clustering process, replace missing s_j column value with m_j column value and find K membership value for s_i.

$$u_{ij} = \frac{1}{\sum_{k=1}^{C} \left(\frac{\|x_i - m_j\|}{\|x_i - m_k\|} \right)^{\frac{2}{v-1}}}$$

Step 4: In Fuzzy C-Means imputation method, object with missing attributes are imputed based on the information about membership degrees and the values of cluster centroids.

$$s_{i,j} = \sum_{k=1}^{c} u_{i,k} * m_{k,j}$$

Step 5: Repeat from the step 3 until to complete the imputation process for all objects in MData.

Step 6: Stop.

In Fuzzy C-Means Parameter based imputation algorithms, the imputation of the missing value is based on the information of the membership degrees and the values of the cluster centroids. The Eq. 6 is used to impute the missing attribute by Fuzzy C-Means Parameter based imputation algorithm [9, 12, 18, 19].

$$x_{i,j} = \sum_{k=1}^{K} U(x_i, C_k) * m_{k,j} \qquad (6)$$

In Fuzzy C-Means clustering algorithm, each object is present in all the clusters by membership value. The object importance to the cluster is identified based on the membership value. The equation shows summation of the product value of the centroid and the membership values of the corresponding object with missing attributes. In Fuzzy Clustering, for a non-reference object x_i which is present in all the cluster and the contribution information to all the cluster is getting from the membership degree. By Fuzzy C-Means Parameter based Imputation algorithm, Contribution of an object to a cluster is identified by the above method, then the knowledge of the cluster is added to impute the missing attribute.

Rough K-Means Parameter Based Imputation Algorithm.

Algorithm : Rough K-Means Parameter based Imputation Algorithm [13].
Input   : Dataset X of n objects with missing attribute values, threshold, w_l, w_u and number of clusters K, $(K<n)$
Output   : Objects without missing values (Imputed Dataset)

Step 0: Given dataset of n objects $X = \{x_1, x_2, ..., x_n\}$, CData $\subset$ X, MData $\subset$ X, where n is the number of objects, CData contain objects with complete data and MData contains objects with missing attribute values. Where MData=$\{s_i \mid i=1,..., T\}$, T is the number of objects in MData.

Step 1: Initialize the number of cluster K.

Step 2: Apply Rough K-Means on CData and Centroid calculation for Rough K-Means Clustering algorithm follows.

$$\vec{m}_k = w_l \sum_{\vec{X}_n \in \underline{C_k}} \frac{\vec{X}_n}{|\underline{C_k}|} + w_u \sum_{\vec{X}_n \in \overline{C_k}} \frac{\vec{X}_n}{|\overline{C_k}|}$$

Step 3: To impute the missing value find the distance between all the approximation and the object with missing attribute MData, minimum distance mean value is impute to the non-reference attribute.

Step 4: Repeat from the step 3 until to complete the imputation all object in MData.

Step 5: Stop.

In Rough K-Means Parameter based imputation algorithm, the rough clustering is applied to the dataset [10, 11, 15–17].

$$x_i = \begin{cases} \frac{\sum_{x_i \in \underline{A}(C_k)} x_i}{|x_j|} \times W_{lower} + \frac{\sum_{x_i \in \bar{A}(C_k) - \underline{A}(C_k)} x_j}{|x_j|} \times w_{upper}, \\ \text{if} \quad x_i \in \underline{A}(C_k) \quad \text{for any} \quad 1 \leq k \geq K, \quad \text{and} \quad x_j \quad \text{is a complete object}, \quad (7) \\ \frac{\sum_{x_j \in \bar{A}(C_k)} x_j}{|x_j|}, \quad \text{if } x_i \notin \underline{A}(C_{k'}) \text{ for all } 1 \leq k' \leq K. \end{cases}$$

The Eq. 7 shows that the imputation is based on the information on a lower approximation object and upper approximation objects. Object with missing value is present in a lower approximation mean, then the information about the lower approximation is used to impute the attribute value. If the non-reference object is present in the upper approximation, than the information about the upper approximation object is used to impute missing values. If the data set with high variance, Rough K-Means Parameter based Imputation that gives optimal accuracy to the imputed value.

4 Application

4.1 Classification

In this study, Back Propagation based Neural Network classification is used to evaluate the accuracy of the imputed value. The importance to choose the neural network based classification has been addressed on the following. Artificial Neural Network is simply referred as Neural Network, is the models of the biological Neurons System (Human brain). The significance of the biological neural systems is parallel distributed processing, neurons are highly interconnected and computation or decision making is very fast compared to any other system. The huge amount of neurons and interconnection between neurons makes the system to get optimal result. The same process of the biological neural system concepts is imitated to build a new computation system called Artificial Neural Network. It contains an input layer, hidden layer and output layer. In input layer each neuron gets signal from p piece of signal and each neuron in input layer that are connected to all other neurons in hidden layers, this way of connection is called a fully connected network. The number of neurons in the hidden layer is chosen based on the application. The transfer function of neuron is chosen based on the application and basically the transfer function used in a neural network are sigmoidal [20–22].

4.2 Medical Dataset

The K-Means, Fuzzy C-Means and Rough K-Means clustering based missing value imputation method is applied to two different medical data sets. The datasets are downloaded from UCI Repository [6]. Table 1 shows the full details of the data sets.

Table 1. Medical datasets.

Dataset name	Total number of instances	Total number of features
Lung Cancer	32	56
Cleveland Heart	303	13

Table 2. Comparison between actual value and the predicted value by various clustering methods

Name of the dataset	No of cluster	Actual places			Imputed value					
					Centroid based			Parameter based		
		Row	Column	Actual values	K-Means	Fuzzy C-Means	Rough K-Means	K-Means	Fuzzy C-Means	Rough K-Means
Lung Cancer	2	3	4	2	1.428571	1.35286	1.8125	2	1.35294	1.33333
		6	6	3	2.5	2.47074	2.15	3	2.47058	2.53333
		11	19	0	0.142857	0.35333	1.5125	2	0.35294	0.26666
		12	20	2	0.642857	0.88165	0.24375	2	0.88235	0.86666
		13	29	2	2.142857	2.17644	2.7375	2	2.17647	2.06666
		20	40	1	2.666666	2.11777	2.7375	2	2.11764	2.06666
		22	43	1	2	2.05888	2	2	2.05882	2
		23	46	3	2	2.05888	2.71875	2	2.05882	2
		24	53	2	2.333333	1.88241	2.6625	3	1.88235	1.8
		27	55	2	1.785714	1.823457	1.24375	2	1.823529	1.866667
	3	3	4	2	1.6	1.35287	1.651	2	1.352941	1.357143
		6	6	3	2.3	2.470825	2.616	3	2.470588	2.571429
		11	19	0	0.2	0.353319	1.52	0	0.352941	0.285714
		12	20	2	0	0.881318	0.26	0	0.882353	0.928571
		13	29	2	2.25	2.176435	2.02	2	2.176471	2.071429
		20	40	1	2.666667	2.117772	2.74	3	2.117647	2.071429
		22	43	1	2	2.058877	2	2	2.058824	2
		23	46	3	2	2.058877	2.72	2	2.058824	2
		24	53	2	2.333333	1.882412	2.68	3	1.882353	1.857143
		27	55	2	1.7	1.823426	1.24	2	1.823529	1.857143
	4	3	4	2	1.75	1.35283	1.80714	2	1.352941	1.25
		6	6	3	2.5	2.47096	2.67142	3	2.47058	2.66666
		11	19	0	1.33333	0.35350	0.08571	2	0.35294	0.33333
		12	20	2	1.11E-16	0.88084	1.67857	0	0.88235	0.83333
		13	29	2	2.66666	2.17644	2.02142	3	2.17647	2.16666
		20	40	1	2.66666	2.11783	1.02142	3	2.11764	2.08333
		22	43	1	2	2.05890	2	2	2.05882	2
		23	46	3	2.33333	2.05890	2.72307	2	2.05882	2
		24	53	2	2.33333	1.88244	2.07692	3	1.88235	1.83333
		27	55	2	1.85714	1.82339	1.9571	2	1.82352	2

The missing attribute are removed and formed the complete data matrix. From the complete data Matrix Manually make the missing Values, the actual value and position of the missing attribute is known. The various clusters based imputation algorithm is applied to the complete data. The Table 2 shows the actual value and the predicted values.

4.3 Application on Lung Cancer Dataset

Table 2 shows the predicted values by six clustering based imputation methods. Overall performance of Rough K-Means Centroid based Imputation method gives optimal results compare to other imputation algorithm for this dataset. Figure 1 shows the Performance of the Imputation algorithms to the Lung Cancer dataset. The Rough

K-Means Centroid based imputation algorithm gives minimized cross-entropy results, than other clustering based imputation algorithms. It shows the correct prediction of missing values.

4.4   Application on Cleveland Heart Dataset

Table 3 shows the predicted values of Cleveland Heart Dataset by six clustering based imputation methods. Compare to other imputation algorithm the proposed Rough K-Means based imputation algorithm gives best results. Figure 2 shows the Performance of the Imputation algorithms of the Cleveland Heart dataset. For this dataset also, the Rough K-Means Centroid based imputation algorithm gives minimized cross-entropy results. That shows the missing values are correctly predicted and placed in the exact clusters.

Table 3. Comparison between actual value and the predicted value by various clustering methods

Name of the dataset	No of cluster	Actual places			Imputed value					
					Centroid based			Parameter based		
		Row	Column	Actual values	K-Means	Fuzzy C-Means	Rough K-Means	K-Means	Fuzzy C-Means	Rough K-Means
Cleveland Heart	2	29	1	43	53.226	53.0676	50.2328	60	54.399	56.6379
		38	4	150	135.38	134.920	132.33	130	134.18	131.931
		82	5	264	299.26	293.651	300.557	305	502.61	267.810
		159	6	0	0.1415	0.15559	0.12662	0	0.1528	0.12069
		186	7	0	0.9171	0.87222	0.59873	2	1.0166	1.17241
		242	8	96	152.19	152.335	147.883	140	152.08	152.662
		269	10	1.8	1.1349	1.06086	1.06592	1.8	1.0585	1.05942
		277	11	2	1.6037	1.58489	1.67754	3	1.58631	1.56521
		278	12	0	0.5801	0.57005	0.57402	2	0.67497	0.70689
		296	13	7	4.6906	4.71292	5.72232	3	4.69786	4.69375
	4	29	1	43	50.934	51.5497	50.7627	60	54.3269	53.4705
		38	4	150	135.4	132.024	145.262	150	132.409	140.314
		82	5	264	191.52	196.658	196.229	197	878.340	184.963
		159	6	0	0.175	0.15986	0.14715	0	0.15334	0.22857
		186	7	0	1.325	1.2425	1.02490	2	1.06715	1.54285
		242	8	96	141.94	144.337	141.431	140	159.212	147.518
		269	10	1.8	1.0477	1.00489	1.03692	1.8	1.03355	0.99591
		277	11	2	1.4545	1.46868	1.81770	3	1.50936	1.46938
		278	12	0	0.5108	0.50534	0.64303	2	0.62602	0.63529
		296	13	7	4.9577	4.91765	6.85117	3	4.79476	4.88888
	6	29	1	43	53.869	52.7962	46.5156	60	54.4027	50.7142
		38	4	150	127.4	129.985	139.246	124	132.307	134
		82	5	264	306.80	258.025	253.708	197	1077.35	248.325
		159	6	0	0.1630	0.16963	0.15520	0	0.15812	0.13513
		186	7	0	1.1333	1.23952	1.16857	2	1.08016	0
		242	8	96	128.43	133.966	120.616	140	196.668	159.529
		269	10	1.8	1.0606	0.8735	0.9785	1.8	1.03556	1.24444
		277	11	2	1.5245	1.4336	1.7681	3	1.4883	1.4444
		278	12	0	0.6304	0.5516	0.7296	2	0.6149	0
		296	13	7	5.3333	5.2220	6.284	3	4.7929	5.3142

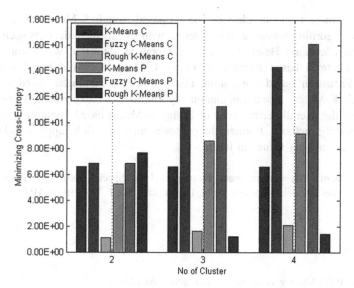

Fig. 1. Performance of imputation algorithms on Lung Cancer dataset

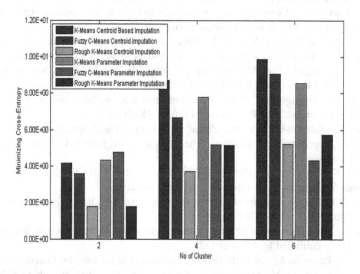

Fig. 2. Performance of imputation algorithms on Cleveland Heart dataset

5 Conclusion

Incomplete data is common in real word data. Analysis of missing values is an important task in statistical model and also increasing in many other fields. The Rough set theory handles the inconsistence and uncertainty on the object. Based on the rough set theory, this paper proposed Rough K-Means Centroid based Imputation. Rough K-Means based Imputation Algorithm overcome the problem of crispness by placing an object to more than one cluster. The proposed Rough K-Means Centroid based Imputation Algorithm was successfully tested with two medical datasets such as Lung Cancer and Cleveland Heart. Further, the proposed algorithm was compared with K-Means Centroid based Imputation, Fuzzy C-Means Centroid based Imputation, K-Means Parameter based Imputation, Fuzzy C-Means Parameter based Imputation and Rough K-Means based Imputation algorithms. The experimental results also showed that the overall performance of Rough K-Means based Imputation was better than the existing methods. Further, the proposed method will be applied to large data sets to handle missing values in future.

Acknowledgment. The present work is supported by Special Assistance Programme of University Grants Commission, New Delhi, India (Grant No. F.3-50/2011(SAP-II)).

References

1. Allison, P.D.: Missing Data. Sage, Thousand Oaks (2001)
2. Gajawada, S., Toshniwal, D.: Missing value imputation method based on clustering and nearest neighbours. Int. J. Future Comput. Commun. **1**(2), 206 (2012)
3. Nelwamondo F.V.: Computational intelligence techniques for missing data imputation. University of the Witwatersrand, Johannesburg
4. Zhang, S., Zhang, J., Zhu, X., Qin, Y., Zhang, C.: Missing value imputation based on data clustering. In: Australian Large ARC Grants China NSF Major Research Program
5. Han, J., Kamber, M.: Data Mining Concepts and Techniques, 2nd edn. Morgan Kaufmann Publishers, San Francisco (2006). ISBN 1-55860-901-6
6. Suguna, N., Thanushkodi, K.G.: Predicting missing attribute values using k-means clustering. J. Comput. Sci. **7**(2), 216–224 (2011). ISBN 1549-3636
7. Kondo, Y., Salibian-Barrera, M., Zamar, R.: A robust and sparse K-means clustering algorithm. Department of Statistics, The University of British Columbia, Vancouver, Canada, January 2012
8. Pavan, K.K., Appa Rao, A., Dattatreya Rao, A.V., Sridhar, G.R.: Single pass seed selection algorithm for k-means. J. Comput. Sci. **6**(1), 60–66 (2010)
9. Havens, T.C.: Fuzzy c-means algorithms for very large data. IEEE Trans. Fuzzy Syst. **20**, 1130–1146 (2012). 1063-6706/$31.00
10. Peters, G.: Some refinements of rough k-means clustering. Pattern Recogn. Soc. **39**, 1481–1491 (2006). Published by Elsevier Ltd.
11. Peters, G., Lampart, M.: A partitive rough clustering algorithm. In: Greco, S., Hata, Y., Hirano, S., Inuiguchi, M., Miyamoto, S., Nguyen, H.S., Słowiński, R. (eds.) RSCTC 2006. LNCS, vol. 4259, pp. 657–666. Springer, Heidelberg (2006). doi:10.1007/11908029_68
12. Panda, S., Sahu, S., Jena, P., Chattopadhyay, S.: Comparing fuzzy-C means and K-means clustering techniques: a comprehensive study. In: Wyld, D.C., Zizka, J., Nagamalai, D. (eds.) ICCSEA 2012. AISC, vol. 166, pp. 451–460. Springer, Heidelberg (2012). doi:10.1007/978-3-642-30157-5_45
13. Li, D., Deogun, J., Spaulding, W., Shuart, B.: Dealing with missing data: algorithms based on fuzzy set and rough set theories. In: Peters, J.F., Skowron, A. (eds.) Transactions on Rough Sets IV. LNCS, vol. 3700, pp. 37–57. Springer, Heidelberg (2005). doi:10.1007/11574798_3

14. Li, D., Deogun, J., Spaulding, W., Shuart, B.: Towards missing data imputation: a study of fuzzy k-means clustering method. In: Tsumoto, S., Słowiński, R., Komorowski, J., Grzymała-Busse, J.W. (eds.) RSCTC 2004. LNCS, vol. 3066, pp. 573–579. Springer, Heidelberg (2004). doi:10.1007/978-3-540-25929-9_70

15. Peters, G., Crespo, F.: An Illustrative comparison of rough k-Means. In: Ciucci, D., Inuiguchi, M., Yao, Y., Ślęzak, D., Wang, G. (eds.) RSFDGrC 2013. LNCS, vol. 8170, pp. 337–344. Springer, Heidelberg (2013). doi:10.1007/978-3-642-41218-9_36

16. Peters, G., Lampart, M., Weber, R.: Evolutionary rough k-medoid clustering. In: Peters, J.F., Skowron, A. (eds.) Transactions on Rough Sets VIII. LNCS, vol. 5084, pp. 289–306. Springer, Heidelberg (2008). doi:10.1007/978-3-540-85064-9_13

17. Lingras, P., Peters, G.: Rough Clustering, vol. 1. Wiley, Hoboken (2011)

18. Bezdek, J.C., Ehrlich, R., Full, W.: FCM: the fuzzy c-means clustering algorithm. Comput. Geosci. **10**(2–3), 191–203 (1984). Pergamon Press Ltd.

19. Dun, J.C.: A fuzzy relative of the ISODATA process and its use in detecting compact well-separated clusters. J. Cybern. **3**(3), 32–57 (1974). Department of Theoretical and Applied Mechanics, Cornell University

20. Reby, D., et al.: Artificial neural networks as a classification method in the behavioural sciences. Behav. Process. **40**, 35–43 (1997). Elsevier Science B.V.

21. Zhang, G.P.: Neural networks for classification: a survey. IEEE Trans. Syst. Man Cybern.—Part C: Appl. Rev. **30**(4), 451–462 (2000)

22. Rey-del-Castillo, P., Cardeñosa, J.: Fuzzy Min–Max Neural Networks for Categorical Data: Application to Missing Data Imputation. Springer, London (2011)

An Optimized Anisotropic Diffusion Approach for Despeckling of SAR Images

Vikrant Bhateja[1(✉)], Aditi Sharma[1], Abhishek Tripathi[1],
Suresh Chandra Satapathy[2], and Dac-Nhuong Le[3]

[1] Department of Electronics and Communication Engineering,
Shri Ramswaroop Memorial Group of Professional Colleges (SRMGPC),
Lucknow 226028, (UP), India
bhateja.vikrant@gmail.com, aditiii065@gmail.com,
abhishekl.srmcem@gmail.com
[2] Department of Computer Science and Engineering,
ANITS, Visakhapatnam, (AP), India
sureshsatapathy@gmail.com
[3] Haiphong University, Haiphong 180000, Vietnam
nhuongld@dhhp.edu.vn

Abstract. This paper focuses to address the suppression of residual speckle content, thereby leading to performance improvement of Anisotropic Diffusion (AD) filtering. The present work proposes an optimized AD filtering approach for despeckling of Synthetic Aperture Radar (SAR) images using Ant Colony Optimization (ACO). The residual speckle suppression has been attained via optimal parameter selection of parameters of AD using ACO algorithm. Further, computation of conductance function via eight-directional gradients (derivatives) leads to effective edge preservation during despeckling. During simulations, the fidelity of restored SAR images is validated using PSNR and SSIM as image quality metrics.

Keywords: Ants · Anisotropic diffusion · Conductance function · Directional gradient

1 Introduction

Remote sensing images are generally the outcome of coherent imaging systems like Synthetic Aperture Radar (SAR), which are generally contaminated with signal-dependent Speckle noise. It is a type of granular noise causing lack of resolution in such images. It is modeled as multiplicative noise and also occurs in images from other modalities like laser illuminated and ultrasound images. It is therefore necessary to apply speckle filtering (coined as 'Despeckling') as an initial step, prior to any feature extraction or high-level processing operation in SAR images. However, such a pre-processing of SAR images is often a compromise between speckle suppression and loss of useful information (like linear and curved features, texture, point objects etc.) [1, 2]. Speckle filtering procedure has been considered to be a non-trivial operation, owing to non-linear relationship between speckled and noise-free images. A common approach to deal with such multiplicative noise models is to convert the same to additive

© Springer Nature Singapore Pte Ltd. 2016
S. Subramanian et al. (Eds.): CSI 2016, CCIS 679, pp. 134–140, 2016.
DOI: 10.1007/978-981-10-3274-5_11

models. The same has been achieved by employing homomorphic transformation by taking logarithm of the SAR images. However, such an approach tends to introduce bias in the filtered image. As an outcome of logarithmic transformation unbiased estimation is coherently mapped to a biased one in spatial domain; this perceptually, affects the brighter regions of the image causing an effective enhancement in the average gray-levels of the local region (homogeneous region) [3]. With such a constraint, despeckling via incoherent averaging only moderately filters the noise at the expense of degradation in spatial resolution [4]. The review of despeckling approaches broadly categorizes the same as Bayesian and Non-Bayesian approaches [5]. Bayesian approaches include local statistics filtering [6] in spatial domain whereas Homomorphic [7] and Wavelet based multi-resolution filtering [8] in transform domain. The functionality in these approaches could be limited in terms of manual threshold selection and ability to discern homogeneous and non-homogeneous regions. Non-Bayesian approaches on the other-hand consist of Order-Statistics [9, 10], Morphological filtering [11, 12] and AD filtering [12–14]. Amongst the Non-Bayesian despeckling approaches, AD filtering has gained ample prominence; wherein the outcome of conduction function(s) in the small regions (within an image) provides effective criteria to discern the image into hetero-geneous and homogeneous regions. Further, the computation of directional gradient serves to discriminate true and false edges [12, 15]. The research in the domain of AD filtering has been evolutionary, starting from the traditional Perona-Malik Anisotropic Diffusion (PMAD) filtering [13], Speckle Reducing Anisotropic Diffusion (SRAD) [16], DPAD [17] and OSRAD [18]. However, the performance deteriorates at higher degree of speckle variances. In addition, the persistence of residual speckle content has been a prime concern in these approaches. This very issue has been focused in this paper to carry out performance improvement of Anisotropic Diffusion (AD) filtering. The present work proposes an optimized AD filtering approach for despeckling of Synthetic Aperture Radar (SAR) images using Ant Colony Optimization (ACO). The residual speckle suppression has been attained via optimal parameter selection of parameters of AD using ACO algorithm. After the introduction and problem definition in Sect. 1; the proposed optimized despeckling approach has been narrated in Sect. 2. Further, Sect. 3 presents the simulation results and Sect. 4 concludes the manuscript.

2 Proposed Optimized Despeckling Approach

The classical isotropic diffusion equation was replaced with the AD differential equation and presented in discrete form as stated below:

$$I_s^{t+1} = I_s^t + \frac{\lambda}{|\eta_s|} \sum_{p \in \eta_s} g(\nabla I_{s,p}) \nabla I_{s,p} \tag{1}$$

The terms of Eq. (1) are defined as in PMAD filtering algorithm [13]. The conductance function of PMAD has been defined in terms of image gradient in Eq. (2).

$$c(x, y, t) = \frac{1}{1 + \left(\frac{\|\nabla I\|}{k}\right)^2} \tag{2}$$

Where: k denotes diffusion constant. The conductance function is selected in such a manner to prevent diffusion across the edges while the diffusion process may continue in homogenous region. The computation of gradient can be made in terms of directional derivatives of each pixel within 3×3 spatial mask centered at any pixel location I(i, j). This can be mathematically expressed in eight directions covering vertical, horizontal and diagonal edges respectively as in Eq. (3).

$$\begin{cases} \nabla_N I_{i,j}^n = I_{i-1,j}^n - I_{i-1,j}^n \\ \nabla_{NE} I_{i,j}^n = I_{i-1,j+1}^n - I_{i,j}^n \\ \nabla_{NW} I_{i,j}^n = I_{i-1,j+1}^n - I_{i,j}^n \\ \nabla_E I_{i,j}^n = I_{i,j+1}^n - I_{i,j}^n \\ \nabla_W I_{i,j}^n = I_{i,j-1}^n - I_{i,j}^n \\ \nabla_{SE} I_{i,j}^n = I_{i+1,j-1}^n - I_{i,j}^n \\ \nabla_{SW} I_{i,j}^n = I_{i+1,j+1}^n - I_{i,j}^n \\ \nabla_S I_{i,j}^n = I_{i+1,j}^n - I_{i,j}^n \end{cases} \tag{3}$$

Hence, the restored pixel can be stated using Eq. (4).

$$I_{i,j}^{n+1} = I_{i,j}^n + \lambda \left[\begin{array}{l} c_N . \nabla_N I_{i,j}^n + c_{NE} . \nabla_{NE} I_{i,j}^n + c_{NW} . \nabla_{NW} I_{i,j}^n + c_E . \nabla_E I_{i,j}^n \\ + c_W . \nabla_W I_{i,j}^n + c_{SW} . \nabla_{SW} I_{i,j}^n + c_{SE} . \nabla_{SE} I_{i,j}^n + c_S . \nabla_S I_{i,j}^n \end{array} \right] \tag{4}$$

Where: the parameter $\lambda \in [0, 1/4]$; and c denotes the diffusion coefficient computed in each direction with necessary directional subscripts as defined in [19]. The factor λ of Eq. (4) serves to exercise control over smoothing action of filter along with edge preservation. In this paper, the optimal selection of value for parameter λ has been performed using Ant Colony Optimization (ACO) algorithm (which is presented below in generalized form as Algorithm-1). This algorithm is based on ant's behavior and their tendency to search for food within shortest path [20–22]. This ACO algorithm has been deployed in this work for optimal parameter selection of AD speckle suppression algorithm.

ALGORITHM 1. ACO for optimal parameter selection of AD speckle suppression

BEGIN
    INITIALIZATION:
        Algorithm parameters: λ, ρ
        Ant population size: K.
        Maximum number of iteration: N_{Max}.
    GENERATION:
        Generating the pheromone matrix trails for the Ant_k.
        Update the pheromone values ;
        $i=1$.
    REPEAT
        FOR $k = 1$ TO K DO
            Computing the cost function for the Ant_k by the formula Eq. (7)
            Computing probability move of ant individual by the formula Eq. (5)
            IF $f(k) < f(x^*)$ THEN
                Update the pheromone values by the formula EQ. (8)
            Set $x^*=k$.
            ENDIF
        ENDFOR
        UNTIL $i > N_{Max}$
END.

This factor λ has been updated based on the movement of ants and pheromone concept. Therefore, within a spatial mask ($w \times w$), the probability of movement of ants based on pheromone trail can be determined as:

$$p_{ij} = \frac{w(i,j).n_{ij}^{0.1}}{\sum\limits_{i=1}^{m}\sum\limits_{j=1}^{n} w(i,j).n_{ij}^{0.1}}$$ (5)

This parameter quantity of pheromone (n_{ij}) has been mapped to mean gradient in a particular direction as:

$$n_{ij} = \frac{[((w(i-1,j) - w(i,j)) + (w(i,j-1) - w(i,j)]}{\sum\limits_{i} w}$$ (6)

The fitness function of ACO may be stated as:

$$\lambda_0 = \lambda + t_i(0.015)$$ (7)

If condition: $p_{ij} > 0$; Eq. (7) is executed otherwise, the spatial mask is shifted to next pixel. Henceforth, λ has been optimized based on the update function for the same using ACO is given by:

$$t_i = (1 - \rho)t_i + \rho(\Delta t_{i,j})$$ (8)

Where: ρ is tuning parameter and the $\Delta t_{i,j}$ denotes the initial pheromone value. The new value of λ is used in Eq. (8) for generation of the restored image. Henceforth, Mean Squared Error (MSE) has been computed taking last processed image as reference and restored image is the one obtained using new value of λ.

$$MSE = \frac{1}{MN}\sum\limits_{i=1}^{M}\sum\limits_{j=1}^{N}\left(x_{i,j} - y_{i,j}\right)^2$$ (9)

The entire algorithmic process is repeated unless the MSE_{i-1} is greater than the value of current MSE_i.

3   Simulation Results and Discussions

The simulation results in this paper are demonstrated on SAR Image of 'Moon Crater' in its normalized form. Results are computed and analyzed on this input test image with different levels of speckle noise from 0.01 to 0.1 variance levels. The spatial mask deployed is of size ($w = 3 \times 3$), ACO parameters ρ is chosen as 0.05 and $\Delta t_{i,j}$ as 0.001. The response of the proposed optimized despeckling algorithm has been demonstrated on speckled image with variance of 0.04 as shown in Fig. 1.

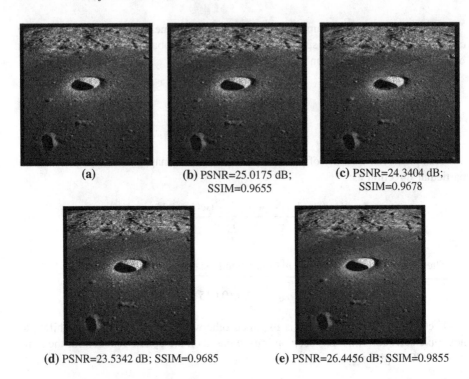

(a)

(b) PSNR=25.0175 dB;
SSIM=0.9655

(c) PSNR=24.3404 dB;
SSIM=0.9678

(d) PSNR=23.5342 dB; SSIM=0.9685

(e) PSNR=26.4456 dB; SSIM=0.9855

Fig. 1. Response comparison despeckling techniques. (a) Noisy SAR (Moon Crater) with variance 0.04. Simulation results obtained using (b) PMAD [13], (c) SRAD [16], (d) OSRAD [18], (e) Proposed optimized despeckling approach.

It has been also compared with other state-of-art despeckling approaches like PMAD [13], SRAD [16] and OSRAD [18] respectively (in same Fig. 1). The objective image fidelity assessment and performance comparison with other approaches has been carried out using Peak Signal to Noise Ratio (PSNR) and Structural Similarity (SSIM) as fidelity metrics [23, 24]. The computed values of aforesaid metrics are also included in Fig. 1. It is evident from Fig. 1(e) that the restored SAR image with optimized despeckling approach shows smoothening in homogenous regions and also preservation of structure and edges significantly. The performance improvement with respect to PMAD has been possible with optimal selection of parameter λ, thereby suppressing

Table 1. Performance evaluation and comparison of proposed optimized despeckling approach with other despeckling approaches using PSNR (dB).

Speckle variance	PMAD [13]	OSRAD [18]	Proposed optimized despeckling approach
0.01	25.6899	25.5522	27.3848
0.04	25.018	23.5342	26.4456
0.06	23.5858	22.2660	24.6192
0.08	22.4178	21.2599	23.0123

the residual speckle content. The performance evaluation using PSNR (dB) at different speckle variance levels has been shown in Table 1.

4 Conclusion

It is known that SAR images consist of sharp edges and variant textures owing to variable landscape. This poses difficulties during speckle filtering in terms of estimation of details and other information content. The present work addresses the issue of residual speckle suppression during filtering action in order to enhance the performance of AD filtering. The same has been achieved using optimal parameter selection of AD filtering using ACO algorithm. It has been observed from simulation results that reduction in residual speckle has been achieved without enhancing computational loads.

References

1. Lu, Y., Gao, Q., Sun, D., Xia, Y., Zhang, D.: SAR speckle reduction using Laplace mixture model and spatial mutual information in the directionlet domain. Neurocomputing **173**, 633–644 (2016)
2. Morandeira, N.S., Grimson, R., Kandus, P.: Assessment of SAR speckle filters in the context of object-based image analysis. Remote Sens. Lett. **7**(2), 150–159 (2016)
3. Chan, Y.K., Koo, V.C.: An introduction to synthetic aperture radar (SAR). J. Prog. Electromag. Res. B **2**, 27–60 (2008)
4. Singh S., Jain A., Bhateja V.: A comparative evaluation of various despeckling algorithms for medical images. In: Proceedings of (ACMICPS) CUBE International Information Technology Conference & Exhibition, Pune, India, pp. 32–37 (2012)
5. Sharma, A., Bhateja, V., Tripathi, A.: An improved Kuan algorithm for despeckling SAR images. Inf. Syst. Des. Intell. Appl. **434**, 663–672 (2016)
6. Bhateja, V., Tripathi, A., Gupta, A.: An improved local statistics filter for denoising of SAR images. In: Proceeding of 2nd International Symposium on Intelligent Informatics (ISI 2013), Mysore, India, vol. 235, pp. 23–29 (2013)
7. Biradar, N., Dewal, M.L., Rohit, M.K.: A novel hybrid homomorphic fuzzy filter for speckle noise reduction. Biomed. Eng. Lett. **4**(2), 176–185 (2014)
8. Bhateja, V., Tripathi, A., Gupta, A., Lay-Ekuakille, A.: Speckle suppression in SAR images employing modified anisotropic diffusion filtering in wavelet domain for environment monitoring. Measurement **74**, 246–254 (2015)
9. Touzi, R.: A review of speckle filtering in the context of estimation theory. IEEE Trans. Geosci. Remote Sens. **40**(11), 2392–2404 (2002)
10. Bhateja, V., Rastogi, K., Verma, A., Malhotra, C.: A non-iterative adaptive median filter for image denoising. In: 2014 International Conference on Signal Processing and Integrated Networks (SPIN), pp. 113–118. IEEE February 2014
11. Lee, I.K., Shamsoddini, A., Li, X., Trinder, J.C., Li, Z.: Extracting hurricane eye morphology from spaceborne SAR images using morphological analysis. ISPRS J. Photogram. Remote Sens. **117**, 115–125 (2016)

12. Argenti, F., Lapini, A., Bianchi, T., Alparone, L.: A tutorial on synthetic aperture radar images. IEEE Geosci. Remote Sens. Mag. **1**(3), 6–35 (2013)
13. Perona, P., Malik, J.: Scale space and edge detection using anisotropic diffusion. IEEE Trans. Pattern Anal. Mach. Intell. **12**(7), 629–639 (1990)
14. Bhateja, V., Singh, G., Srivastava, A., Singh, J.: Despeckling of ultrasound images using non-linear conductance function. In: Proceedings of (IEEE) International Conference of Signal Processing and Integrated Networks (SPIN-2014), Noida (U.P.), India, pp. 722–726 (2014)
15. Bhateja, V., Sharma, A., Tripathi, A., Satapathy, S.C.: Modified non linear diffusion approach for multiplicative noise. In: Proceedings of 5th International Conference on frontiers of Intelligent Computing Theory and Applications (FICTA-2016), pp. 1–8, September 2016
16. Yu, Y., Acton, S.T.: Speckle reducing anisotropic diffusion. IEEE Trans. Image Process. **11** (11), 1260–1270 (2002)
17. Aja-Fernández, S., Alberola-López, C.: On the estimation of the coefficient of variation for anisotropic diffusion speckle filtering. IEEE Trans. Image Process. **15**(9), 2694–2701 (2006)
18. Krissian, K., et al.: Oriented speckle reducing anisotropic diffusion. IEEE Trans. Image Process. **16**(5), 1412–1424 (2007)
19. Gupta, A., Tripathi, A., Bhateja, V.: Despeckling of SAR images via an improved anisotropic diffusion algorithm. In: Satapathy, S.C., Udgata, Siba K., Biswal, B.N. (eds.) FICTA 2013. AISC, vol. 199, pp. 747–754. Springer, Heidelberg (2013). doi:10.1007/978-3-642-35314-7_85
20. Blum, C.: Ant colony optimization: introduction and recent trends. Elsevier Phys. Life Rev. **2**, 353–373 (2005)
21. Marco, D., Birattari, M., Stützle, T.: Ant colony optimization. IEEE Comput. Intell. Mag. **1**(4), 28–39 (2006)
22. Gao, L., Gao, J., Li, J., Plaza, A., Zhuang, L., Sun, X., Zhang, B.: Multiple algorithm integration based on ant colony optimization for end member extraction from hyperspectral imagery. IEEE J. Select. Top. Appl. Earth Obs. Remote Sens. **8**(6), 2569–2582 (2015)
23. Bhateja, V., Misra, M., Urooj, S., Lay-Ekuakille, A.: Bilateral despeckling filter in homogeneity domain for breast ultrasound images. In: Proceedings of 3rd (IEEE) International Conference on Advance in Computing. Communication and Informatics (ICACCI-2014), Greater Noida (U.P.), India, pp. 1027–1032 (2015)
24. Tripathi, A., Bhateja, V., Sharma, A.: Kuan modified anisotropic diffusion approach for speckle filtering. In: Proceedings of (Springer) 1st International Conference on Intelligent Computing and Communication, (ICIC^2-2016), Kalyani (W.B.), India, pp. 01–09, February 2016

Mammogram Classification Using ANFIS with Ant Colony Optimization Based Learning

K. Thangavel[1(✉)] and A. Kaja Mohideen[2]

[1] Department of Computer Science,
Periyar University, Salem 636011, Tamil Nadu, India
drktvelu@yahoo.com
[2] Department of Applied Mathematics and Computational Sciences,
PSG College of Technology, Coimbatore 641004, Tamil Nadu, India
kaja.akm@gmail.com

Abstract. Women Breast Cancer has high incidence rate in worldwide. Computer aided diagnosis helps the radiologist to diagnose and treat the breast cancer at early stage. Recent studies states that Adaptive Neural Fuzzy Inference System (ANFIS) classifier achieves notable performance than the other classifiers. The major forte of ANFIS is that it has the robust learning mechanism with fuzzy data. However, the connections between the layers are not pruned for their significance. An Ant Colony Optimization (ACO) based learning is proposed in this paper to improve ANFIS classifier with a novel pruning strategy. The proposed algorithm is inspired from the social life of a special species called 'Weaver Ants'. The proposed classifier is evaluated with the mammogram images from MIAS database, the quantified results show that this weaver ant based learning strategy improves the ANFIS classifier's performance.

1 Introduction

Breast cancer causes the leading death among women in worldwide. Early diagnosis could help to increase the survival rate. Mammography is one of the most reliable methods available for early detection [1]. At the same time, Computer Aided Diagnosis (CAD) plays a major role on providing second opinion for the radiologists, as they are capable to work with huge data and consistency. Numerous CAD systems has been developed and studied in the past decades. Figure 1 shows a typical CAD system that involves 5 major phases such as preprocessing, segmentation, feature extraction, classification and performance evaluation. This paper focuses on the classification phase. Neural Networks (NNs) is one of the most successful algorithm applied for the supervised classification task. The NNs are capable of mapping the spatial or spectral data inputs to output through learning algorithms. However, NNs are not suitable for vague and incomplete data, and they are adaptive enough to derive the associations between numerical samples. Fuzzy Logic (FL) is one such concept could handle such data adaptively using inference rules. But FL is unable to learn from the samples for setting suitable parameters by its own. Then the Fuzzy Neural Networks (FNN) have been introduced [2], a hybrid solution, which is able to construct the fuzzy inference rules through learning. The Takagi-Sugeno fuzzy systems are successfully proven to work with any nonlinear function approximation problems [2].

© Springer Nature Singapore Pte Ltd. 2016
S. Subramanian et al. (Eds.): CSI 2016, CCIS 679, pp. 141–152, 2016.
DOI: 10.1007/978-981-10-3274-5_12

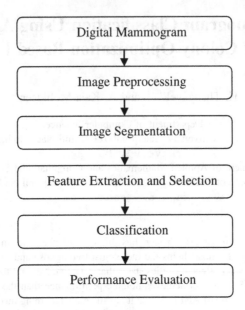

Fig. 1. Typical CAD system framework

Mitra and Hayashi [3] conducted a study of neuro-fuzzy algorithms for real-time applications, and conclude the investigation with the high quality results. The rule-refinement strategy and building adaptive rule concept is explored by Mousavi et al. [4]. Huang et al. [5] proposed an integrated algorithm for medical diagnosis using Adaptive Neuro Fuzzy Inference System (ANFIS) with two-pass learning algorithms. The ANFIS produces greater results for the computer aided diagnosis of glaucomatous. Walia et al. [6] implemented ANFIS for tuberculosis diagnosis, ends up with better accuracies.

Though the ANFIS outperforms NN performance, they are not suitable for the problems with higher dimension of inputs. The increase in dimension will increase the fuzzy rules exponentially, which requires high computation cost and it might affects the classification performance. To resolve this problem, a novel Ant Colony Optimization (ACO) based learning is proposed in this paper.

Ant Colony Optimization (ACO) was introduced by Dorigo colleagues in 1990s, as a novel nature-inspired metaheuristic for the solution of Combinatorial Optimization (CO) problems [7]. ACO algorithm simulates the foraging behavior of real ants. The ants starts their travel in random, once an ant find the food source, it carries a piece of food and return back to the nest. The path of the return journey is marked with a chemical substance called pheromone, which is self-evaporating in nature. Another ant starting from the nest will start following this path to reach the food source. This way of indirect communication helps them to find the shortest path has the highest pheromone deposits. This behavior is simulated to solve CO problems.

The general procedure for ACO is given below:

   Procedure ACOMetaheuristic

         Set parameters, initialize pheromone trails

         while (termination_condition not met) **do**

            Construct_Ants Solutions

            Apply_Local_Search (optional)

            Update_Pheromones

            Daemon_Actions

       end

   end

In this paper, a novel ACO algorithm is proposed to improve the ANFIS classifier's performance. The ACO algorithm used here is modeled from the behavior of an ant species called 'Weaver Ants', so the algorithm is named as Weaver Ant Colony Optimization (WACO).

The rest of the paper is organized as follows: The following section describes the basic architecture of Sugeno-type ANFIS. Section 3 explains the proposed WACO-based ANFIS classifier. Section 4 presents the experimental setups for the evaluation of the proposed classifier. The quantified results are summarized and explained in Sect. 5. Section 6 concludes the paper.

2 Adaptive Fuzzy Neural Network Inference System

ANFIS is a feed-forward multi-layer classifier, where the learning is achieved by the concept of neural network and the inputs to output mapping is done with the help of fuzzy logic [8]. Figure 2 shows a typical ANFIS architecture, has 5 layers by default. Initially the data samples are fed to the network.

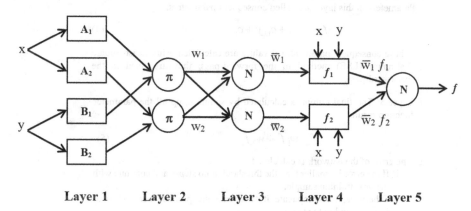

Layer 1   Layer 2   Layer 3   Layer 4   Layer 5

Fig. 2. Structure of a typical ANFIS network

The first layer finds the linguistic descriptions of each input, the number of descriptions are depends upon the type of fuzzy membership functions used. Making use of more number of fuzzified values will increase the complexity of the architecture.

Algorithm. ANFIS

Step 1. For each data sample in the training set

a. Read a data sample and fed to the input layer of ANFIS.

b. Layer-1: A membership function is defined for every input to receive the linguistic descriptions (fuzzified values) of it.

In general, there are five different types of membership functions are used such as bell-shaped, triangular, trapezoidal, Gaussian, sigmoidal and polynomial based functions. This research work uses triangular-shaped membership function as it is consistent and simple to use [9]. The definition for triangular-shaped membership function is given below.

$$\mu_i(x) = \begin{cases} \dfrac{x-a_i}{b_i-a_i} & a_i \le x_i \le b_i \\ \dfrac{c_i-x}{c_i-b_i} & b_i \le x_i \le c_i \\ 0 & Otherwise \end{cases} \quad \text{and} \quad \mu_i(y) = \begin{cases} \dfrac{y-s_i}{t_i-s_i} & s_i \le y_i \le t_i \\ \dfrac{u_i-y}{u_i-t_i} & t_i \le y_i \le u_i \\ 0 & Otherwise \end{cases} \quad (1)$$

Where, x and y, are the input samples, {a, b, c} and {s, t, u} are the premise parameters. Initially these values are assigned in random for the forward-pass, and they are updated in the backward-pass using gradient descent method.

c. Layer-2: Each node measures the firing strength of each rule by finding the product of membership function values from the previous layer.

$$w_i = \mu_{A_i}(x)\,\mu_{B_i}(y), i = 1, 2. \quad (2)$$

d. Layer-3: The normalized firing the strength of each rule is calculated in this layer.

$$\overline{w_i} = \frac{w_i}{w_1 + w_2} \quad (3)$$

e. Layer-4: The nodes compute a parameter function on the layer 3 output. Parameters in this layer are called consequent parameters.

$$f_1 = c_{11}x + c_{12}y + c_{10} \quad (4)$$

These consequent parameter (c) values are calculated using Least-Square Estimator (LSE) method in the forward-pass; they are fixed in the backward-pass.

f. Layer-5: The final output is calculated by summing up all the parameter function results.

$$f = \overline{w_1}f_1 + \overline{w_2}f_2 \quad (5)$$

g. The error of the network is calculated

i. If the error is smaller than the threshold, goto step-a and continue with the next training sample.

ii. Otherwise, backpropagate it to update the premise parameters at Layer-1, and goto step-b.

Step 2. End of Training

Step 3. Test the network model with the samples from test data set.

The second layer finds the product of all the connected membership values. Third layer normalize the incoming values, fourth layer is used for defuzzification and the output is calculated at the fifth layer. ANFIS is used for solving parameter identification problems using a hybrid learning rule combining the back-propagation gradient descent and a least-squares method. The following algorithm details the ANFIS learning procedure. The algorithm has two flows: forward-pass (moving from Layer-1 to Layer-5) to calculate the output of the network, backward-pass for parameter updation (at Layer-4 and Layer-1).

3 The Proposed WACO–ANFIS

The proposed ant algorithm is modeled from the behavior of a special ant species called 'Weaver Ants'. The weaver ants are known for their nest building characteristics. They choose a bunch of living leaves from the branch of a plant, and started pinching and pasting the leaves to construct a closed nest for their living [10]. This behavior of nest construction is simulated to bind the neurons between successive layers in the ANFIS architecture. Each layer of the ANFIS is considered as one leaf, the neurons in the layers are assumed as the pinched made by the weaver ants, and let the ants to connect portion of a leaf from one layer to another, makes the connection between the layers. As discussed in the previous section here we are using a five-layer ANFIS, with a small change in the architecture to form the closed nest. Initially the ants are allowed to select the paths between the layers in random, and the network error is calculated. From the next iteration, the paths are selected based on heuristic search and the error is checked against with the previous error value, if the error gets reduced then the current path's pheromone is updated, otherwise the unselected paths pheromone are updated. This procedure is continued for the entire learning phase.

In general, the neurons from one layer are connected to the neurons in the successive layer only. Here in the proposed WACO-ANFIS adds one more connection from layer-1 to layer-5 as illustrated in Fig. 3.

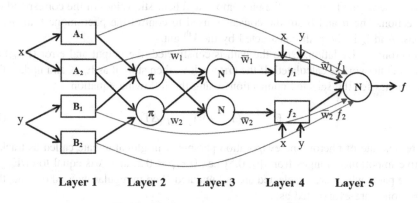

Fig. 3. The proposed ANFIS architecture

The proposed architecture is able to reduce the number of epochs competitively, though it might be a simple modification. Owing to an extra connection is added, the output of the layer-5 is calculated with the modified function as defined below

$$f = \overline{w_1}f_1 + \overline{w_2}f_2 + \frac{\mu_{A_1}}{\mu_{A_2}} + \frac{\mu_{B_1}}{\mu_{B_2}} \tag{6}$$

With this proposed architecture, the WACO based learning approach begins as described in the following text.

Initially 'k' number of ants is chosen to construct the path between the layers. The construction has the assumption that the ants are allowed to move only in the forward direction and the connection should be established only with the successive layers (except for the connection between layers 1 to 5). So, the possible connections are from L1–L2, L2–L3, L3–L4, L4–L5 and L1–L5.

Pheromone Initialization: The connections between the layers are assigned with the initial pheromone value T0, a random value between −0.5 to +0.5 [11]. These pheromone values are maintained in matrices V_{12}, V_{23}, V_{34}, V_{45}, and V_{14}.

For each pheromone matrix, a corresponding flag matrix is used to denote whether the path has been selected or not with the Boolean values 0 or 1 ('0' for unselected and '1' for selected). At first, the error of the network is calculated by keeping all the connections selected and stored in E_{min}.

Path Construction: From the second iteration, let 'k' number of ants to choose the paths and the network error is calculated. Figure 4 illustrates a sample path construction between the layers, the solid lines represents the paths chosen by the ants whereas the dotted lines represent the unselected paths. The paths are chosen based on the following probabilistic transition rule

$$P_{ij}^k(t) = \begin{cases} \dfrac{[T_{ij}(t)]^{\alpha}[\eta_{ij}(t)]^{\beta}}{\sum\limits_{u \in j^k}[T_{ij}(t)]^{\alpha}[\eta_{ij}(t)]^{\beta}} & i \in j^k \\ 0 & Otherwise \end{cases} \tag{7}$$

Where $P_{ij}^k(t)$ is the probability of the connection between layers L_i and L_j, for the kth ant at 't' iteration. T and η are the pheromone and heuristic values of the corresponding connection. The α and β are the constants used to control the pheromone trail oscillations. And j_k is the path constructed by the kth ant.

Pheromone Updation: Once the path is constructed, the output and error (L_{min}) of the neural network is calculated. If the error is low, then store it as E_{min}, and update the pheromones for the selected connections using the following equation

$$[T]^{t+1} = (1 - \rho)[T]^t + \rho[\Delta T] \tag{8}$$

where ρ is rate of pheromone evaporation parameter in global update called as track's relative importance, ranges from [0, 0.5] i.e., $0 < \rho < 0.5$, and Δ is equal to (1/E_{min}).

The paths which are unselected are not updated. And in regular interval of time, the pheromones are evaporated as:

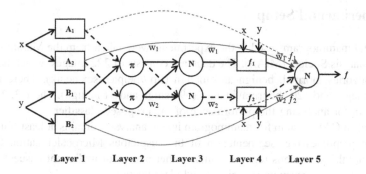

| Layer 1 | Layer 2 | Layer 3 | Layer 4 | Layer 5 |

Fig. 4. Path construction in weaver ant based learning

$$[T]^{t+1} = (1 - \rho)[T]^{t} \tag{9}$$

And for the next iteration, once again a random number of ants are allowed to construct the path and the error is calculated to update the pheromones. This procedure is repeated till the error is less than the threshold. And the network learning is continued with the next pattern to be trained. The algorithm for the proposed Weaver Ant Colony Optimization based Backpropagation Neural Network (WACO-BPN) learning is summarized in Fig. 5.

Inputs: Data Samples, DB_{ij} with the Class Labels C_i
Output: Optimized ANFIS Architecture.
a.   Choose the number of neurons in L1, L2, L3, L4 & L5.
b.   Construct a WACO-ANFIS network.
c.   Choose the Training Samples.
d.   Initialize the pheromones for the connections
e.   For each training sample
    i.   Let 'k' number of ants to choose the connections.
    ii.   Forward-Pass: With the selected connections, find the outputs at L2, L3. And find the consequent parameter values using LSE method, and find the outputs at L4 and L5.
    iii.   Calculate the network error.
    iv.   Backward-Pass: Using gradient descent method, update the premise parameters.
    v.   Update the pheromones based on the network output.
    vi.   If the error is greater than the threshold, continue at step-ii, otherwise continue with step-i with the next training sample.

Fig. 5. WACO-BPN algorithm

4 Experimental Setup

The digital mammograms used in the experiments were taken from the Mammographic Image Analysis Society (MIAS). The database consists of 322 images, which belong to three categories: normal, benign and malign (ftp://peipa.essex.ac.uk). There are 202 normal images and 120 abnormal images (69 benign and 51 malign). These 322 images are used in for analyzing the performance of the proposed classifier.

A typical CAD system for mammogram image analysis contains at least four stages including preprocessing, segmentation of the suspicious Microcalcification Clusters (MCCs, are the suspicious regions contain either benign or malignant tissues), feature extraction from the suspicious regions, and classification.

Initially the mammogram images are enhanced by removing the x-ray labels and pectoral muscle region using simple binary thresholding method [12]. Then the mammogram images are subjected to contrast enhancement using a Walking Ant Histogram Equalization [13]. Next, an improved watershed transformation is applied to derive the initial segmentation, and a Guided Ant Colony Algorithm (GACA) has been employed to merge the regions based on their homogeneity, into two clusters such as: normal and abnormal. The homogeneity is measured with Local Spectral Histogram (LSH) to analyze the normal and abnormal region [14].

From the segmented mass regions, both the textural and shape features are extracted. For the textural features, the Spatial Gray-Level Dependence (SGLD) Method is used to construct the co-occurrence matrix and 14 Haralick features have been extracted. And for each mass region, 7 signal contrast and 13 shape features are extracted [15, 16]. Totally 34 features are calculated for each mass region as shown in Table 1, and the most relevant features are selected using a Leaf-cutter Ant Colony Optimization [17].

Table 1. Features extracted from microcalcification clusters

Feature types	Extracted features
Textural features	Angular second moment, contrast, correlation, variance, inverse difference moment, sum average, sum variance, sum entropy, entropy, difference variance, difference entropy, information measure of correlation I & II, and maximal correlation coefficient
Signal contrast	Minimum, maximum, mean, median, standard deviation, skewness, kurtosis
Shape features	Area, convex area, filled area, perimeter, orientation, eccentricity, euler number, roundness, EquivDiameter, solidity, extent, Centroidx, Cedntroidy

The reduced feature set contains two texture features, (Contrast, Sum Entropy) two contrast features (Skewness, Kurtosis) and one shape feature (Roundness). Further the abnormal images are classified as given in Table 2 based on mass types and tissue types, these dataset are used for evaluating the proposed WACO-ANFIS training algorithm under these class labels.

Table 2. Summary of mammogram image dataset

Types of mass	Types of breast tissue			Total
	Dense	Fatty	Glandular	
Architectural distortion	7	6	6	19
Asymmetry	7	4	4	15
Calcification	10	6	7	23
Circumscribed	3	12	8	23
Ill-defined	2	7	5	14
Spiculated	7	5	7	19
Total	36	40	37	113

According to the annotations and classifying the masses in the following sizes: small lesions have the pixel radius ≤40, medium size has the radius ≤80 and above 80 pixel radius lesions are considered as large in size, there were 53, 42 and 18 masses in each interval, respectively.

The performance of the proposed ANFIS training is compared with the conventional ANFIS [2], Particle Swarm Optimization based ANFIS (PSO-ANFIS) [18], Rough Set based ANFIS (RS-ANFIS) [18], and with our recently proposed classification algorithm which hybrids WACO with BPN [19]. As mentioned in Table 2, here we have four groups of classifications for the feature patterns, (i) major classification as benign or malign, (ii) based on mass types, (iii) based on tissue types, and (iv) based on lesion size. The classifier is evaluated on each of the four classes independently and the performance is analyzed with the four metrics: True Positive (TP), True Negative (TN), False Positive (FP) and False Negative (FN). Where TP and TN as the numbers of correctly classified positive and negative samples, FP and FN for the numbers of incorrectly classified positive and negative samples, i.e. false alarms and missed positives. Several metrics can be determined below for quantitative evaluations:

$$\text{True Positive Rate (TPR)} = TP/(TP + FN) \tag{10}$$

$$\text{False Positive Rate (FPR)} = FP/(FP + TN) \tag{11}$$

5 Results and Discussions

In this section, comprehensive experiments are conducted and the results are presented. Quantitative evaluations are used to validate the effectiveness of our proposed method. Once the mammograms that contain masses have been detected, the algorithms have to be capable of precisely identifying the position and borders of them. This capability is evaluated using ROC analysis, with the emphasis on performance with respect to the different morphological aspects as given in MIAS annotations: the mass type, lesion size and breast tissue type. For this classification scheme, the proposed WACO-ANFIS learning approach obtains the highest accuracy (98%). Next to this, the PSO-ANFIS and RS-ANFIS and WACO-BPN approaches are able to cover 97%, 96% and 97% of

Table 3. Comparison of classifiers' performance

Classifiers	Accuracy			
	Benign vs. malign	Mass types	Tissue types	Lesion size
WACO-ANFIS	98.1321	95.6098	97.7778	91.2361
PSO-ANFIS	97.7611	95.1894	97.0595	90.4231
RS-ANFIS	96.8941	96.1410	96.6559	88.9101
WACO-BPN	97.2207	96.5292	97.3496	89.9701

area under the curve. Table 3 summarizes the quantified results. Figures 6, 7, 8 and 9 shows the ROC curve for the classifiers performance on benign vs. malignant, mass type, tissue types and lesion size based classifications.

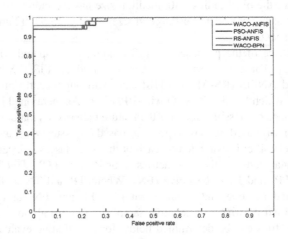

Fig. 6. ROC graph for classifiers performance on benign and malign classification

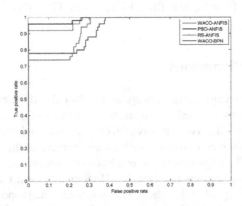

Fig. 7. ROC graph for classifiers performance on mass types classification

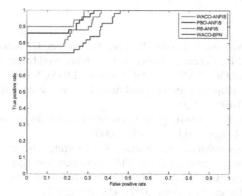

Fig. 8. ROC graph for classifiers performance on tissue types classification

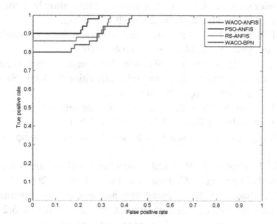

Fig. 9. ROC graph for classifiers performance on lesion sizes classification

6   Conclusions

A novel learning algorithm to improve the ANFIS classifier is proposed in this paper. The connections between the layers in the ANFIS architecture are pruned for their significance using a Weaver Ant Colony Optimization (WACO) algorithm. The proposed classifier is used to classify the mammogram images on four different classification schemes. The performance is compared and analyzed with the well-known conventional ANFIS and the other three methods. The result indicates that the proposed ANFIS outperforms the other.

References

1. Ferlay, J., Soerjomataram, I., Dikshit, R., Eser, S., Mathers, C., Rebelo, M., Parkin, D.M., Forman, D., Bray, F.: Cancer incidence and mortality worldwide: sources, methods and major patterns in GLOBOCAN 2012. Int. J. Cancer **136**(5), E359–E386 (2015)
2. Jang, J.S.: ANFIS: adaptive-network-based fuzzy inference system. IEEE Trans. Syst. Man Cybern. **23**(3), 665–685 (1993)
3. Mitra, S., Hayashi, Y.: Neuro-fuzzy rule generation: survey in soft computing framework. IEEE Trans. Neural Netw. **11**(3), 748–768 (2000)
4. Mousavi, S.J., Ponnambalam, K., Karray, F.: Inferring operating rules for reservoir operations using fuzzy regression and ANFIS. Fuzzy Sets Syst. **158**(10), 1064–1082 (2007)
5. Huang, M.L., Chen, H.Y., Huang, J.J.: Glaucoma detection using adaptive neuro-fuzzy inference system. Expert Syst. Appl. **32**(2), 458–468 (2007)
6. Walia, N., Tiwari, S.K., Malhotra, R.: Design and identification of tuberculosis using fuzzy based decision support system. Advances in CS and IT **2**, 57–62 (2015). ISSN 2393-9915
7. Dorigo, M., Maniezzo, V., Colorni, A.: Ant system: optimization by a colony of cooperating agents. IEEE Trans. Syst. Man Cybern. **26**(1), 29–41 (1996)
8. Wang, Y.M., Elhag, T.M.: An adaptive neuro-fuzzy inference system for bridge risk assessment. Expert Syst. Appl. **34**(4), 3099–3106 (2008)
9. di Sciascio, F., Carelli, R.: Fuzzy basis functions for triangle-shaped membership functions: Universal approximation-MISO case. In: 6th International Fuzzy Systems Association World Congress (IFSA 1995), pp. 439–442 (1995)
10. Deneubourg, J.L., Aron, S., Goss, S., Pasteels, J.M.: The self-organizing exploratory pattern of the argentine ant. J. Insect Behav. **3**(2), 159–168 (1990)
11. Pollack, J.B.: Backpropagation is sensitive to initial conditions. Complex Syst. **4**, 269–280 (1990)
12. Mohideen, A.K., Thangavel, K.: Removal of pectoral muscle region in digital mammograms using binary thresholding. Int. J. Comput. Vis. Image Process. **2**(3), 21–29 (2012)
13. Mohideen, A.K., Thangavel, K.: Contrast enhancement of digital mammograms using a novel walking ant histogram equalization. Int. J. Comput. Vis. Robot. **5**(2), 181–201 (2015)
14. Mohideen, A.K., Thangavel, K.: A guided ant colony algorithm for detection of microcalcification clusters. Int. J. Biomed. Eng. Technol. **10**(4), 309–324 (2012)
15. Haralick, R.M., Shanmugam, K.: Textural features for image classification. IEEE Trans. Systems Man Cybern. **3**(6), 610–621 (1973)
16. Conant-Pablos, S.E., Hernández-Cisneros, R.R., Terashima-Marín, H.: Feature selection for the classification of microcalcifications in digital mammograms using genetic algorithms, sequential search and class separability. Genet. Evol. Comput.: Med. Appl. 69–84 (2011)
17. Mohideen, A.K., Thangavel, K.: Leafcutter ant colony optimization algorithm for feature subset selection on classifying digital mammograms. Int. J. Appl. Metaheuristic Comput. **5**(3), 23–43 (2014)
18. Jiang, H., Kwong, C.K., Siu, K.W.M., Liu, Y.: Rough set and PSO-based ANFIS approaches to modeling customer satisfaction for affective product design. Adv. Eng. Inform. **29**(3), 727–738 (2015)
19. Mohideen, A.K., Thangavel, K.: Weaver ant colony optimization-based neural network learning for mammogram classification. Int. J. Swarm Intell. Res. (IJSIR) **4**(3), 22–41 (2013)

Network Computing

Authentication in Wireless Sensor Networks Using Dynamic Identity Based Signatures

S.D. Suganthi[✉], R. Anitha, and P. Thanalakshmi

PSG College of Technology, Coimbatore, India
sd_suganthi@yahoo.com, anitha_nadarajan@mail.psgtech.ac.in,
ptl@amc.psgtech.ac.in

Abstract. A Wireless sensor network (WSN) is composed of a large number of sensor nodes, which perform multiple tasks, namely sensing, data processing and forwarding of observed data. WSNs nodes may possess sensitive data that are prone to various attacks. For such a network to be viable, integrity and authenticity should be provided to the data generated by the sensor nodes. For example, in military surveillance and enemy tracking applications, the localization system of the nodes is the target for many attackers. In such applications, the base station would broadcast the command for localization to all the sensor nodes in the field. The sensor nodes would respond to this query with the required data. Any compromised node at this point would generate false data and may lead to miscalculation of the localization process and incorrect decision making. Hence, a resilient authentication is necessary to authenticate a node. As a first step towards this objective, a lightweight identity based signature for authentication of the sensor nodes is proposed in this paper. The scheme uses "fingerprint", i.e. a lifetime secure memory fraction in the sensor nodes as a parameter for signature generation. In addition, the parameters for fingerprints are generated dynamically and the computed fingerprint values are not stored permanently in the hardware. Because of these features, the sensor nodes can overcome identity based attack like Sybil attack. Also, it is impossible to read the contents of the sensor node even if the node is captured by the attacker. The security proof for this scheme is based on the Computational Diffie-Hellman assumption and proved in the random oracle model. On the computation point of view, the proposed scheme requires minimal operations in signing than the existing identity based signature approaches.

Keywords: Wireless sensor network · Identity based signature · Authentication · Security · Sensor hardware · Bilinear pairing

1 Introduction

Wireless Sensor Networks have emerged as a promising way for communicating and distributing information in the wireless environment. They are widely used in applications like environmental monitoring, disaster handling, traffic control,

© Springer Nature Singapore Pte Ltd. 2016
S. Subramanian et al. (Eds.): CSI 2016, CCIS 679, pp. 155–168, 2016.
DOI: 10.1007/978-981-10-3274-5_13

object tracking, battlefield monitoring and various ubiquitous applications. Due to the hostile nature of the network, the nodes are vulnerable to various physical attacks which further cause node compromise, node cloning, man-in-the-middle attack and replay attack. Hence techniques that would provide a high level of security against these attacks are essential. In applications like surveillance and target tracking, finding the location of the sensor node is crucial which has become the target for the attackers. The cluster head forwards the base stations query to the sensor nodes, which in turn will respond to the query. In such cases, if a node is compromised the location of a node would be miscalculated leading to incorrect decision making. As the data collected by the sensor nodes are valuable and sensitive, an efficient source authentication procedure is required in order to secure the data sent from a sensor node to a cluster head node or to a base station. Though the traditional authentication schemes using hash chains are efficient in terms of processing and energy consumption, they suffer from the following issues:

– Slow speed in large scale sensor networks
– DOS attack against storage due to late authentication
– Not scalable in terms of number of senders

To handle the above mentioned issues, a lightweight identity based digital signature scheme is proposed in this paper.

1.1 Related Work

In WSNs, digital signatures are used for source authentication of messages broadcast from the base station to the sensor nodes. For this purpose, digital signatures are generated by the base station and verified by the sensor nodes. Node to node authentication is a key property to support a number of security functionalities. Any scheme that is used to revoke misbehaving nodes is based on the certainty that node identities are correctly detected. A number of powerful attacks like Sybil attack, where a node illegitimately claims multiple identities, are mitigated if authentication is enforced [8]. μ Tesla [20] is a lightweight symmetric-key based cryptographic primitive designed for efficient authentication of broadcast messages in WSN.

In [2], authors conducted a survey of several authentication schemes used in wireless sensor networks. Time Synchronized μ Tesla, One Time Signature, and Public Key Authentication are used for broadcast authentication. Kerberos is a network authentication system that uses a trusted third party to authenticate two entities (i.e., to prove their identity to one another) by issuing a shared session key between them [8]. Identity based signatures are particularly suitable for power constrained devices such as sensor nodes in a WSN. The public key generated based on the identity provides non-repudiation of sensor data, and the signature authenticates the message sent from the sensor nodes to the base station. Several identity based signature algorithms [2,6,9–11,15,18,22] have been proposed. In [1], the authors use the timestamp of the message and the current time in the node for authentication. The difference between the timestamp in the authentication request message and the current time of the node is compared with a threshold value. Also, the signature verification algorithm makes use of the timestamp that comes with the authentication request message. This requires a tight synchronization between the sender node and the receiver node, which may not be feasible in a real time application.

Inspired by the acceleration technique, Benzaid et al. [4] proposed an accelerated verification of digital signatures generated by vBNN-IBS. In this scheme, the sensor node involves two of its neighboring nodes for intermediate computation of certain parameters in the signature verification to speed up the process. Though the scheme aims to reduce the time for verification, the energy consumption would be high during verification and also the neighboring nodes deplete their energy in this calculation process. In addition, if the neighboring nodes are captured by the attacker, then the verification process may not be successful leading to a Denial of Service attack.

The authors in [21] used a pairing-optimal IBS scheme with message recovery to improve the communication and signature verification costs. In fact, there exists a pairing-optimal IBS scheme proposed by [2], where the resulting signature consists of a single element of the underlying group and a 160-bit hash value at an 80-bit security level. In this scheme, the original message is not required to be transmitted together with signature since it can be recovered during the verification process. In [8], multiple copies of the same keys are distributed in the network which may lead to faking of sensor identities. Chen et al. [7] have used bilinear pairing for authentication between sensor nodes, sensor and the cluster node. This scheme uses HMAC to support authentication between the sensor nodes and the cluster node. Here the BS selects a random number and calculates the sensor node ID and cluster node ID with some random value it hashes. The authentication happening between the cluster node and the sensor node is through the exchange of the nonce values and keyed hash of these nonce concatenated with the authentication request message. The scheme requires the cluster nodes to be equipped with GPS capability.

Sakai and Kasahara [19] proposed an efficient identity based signature scheme. Though the scheme achieves efficiency in computation by reducing the number of pairing operations in verification, it is not secure since universal forgery of the signature of any user is possible. Bellare et al. [3] modified this scheme and proved it to be UF-CMA secure under the CDH assumption in the Random Oracle model. Libert and Quisquater [12] proved that their scheme provides better security guarantees than its counterparts for the same security levels. Later, Barreto et al. [2] designed an identity based signature scheme which is EUF-CMA secure and claimed it to be the most efficient scheme as it uses a single pairing operation for signature verification. Tso et al. [22] proposed an identity based signature scheme with the idea of reducing the signature size. Although the communication cost of this scheme is little larger than that of a short signature, the computational cost is less than that of Boneh et al.'s [5] short signature in the verification phase. Narayan and Parampalli [15] proposed an identity based signature in the standard model with reduced public parameter size. It is secure against a signature forgery attack in the adaptive identity notion of security.

Our Contribution: The main contribution of our paper is an Identity Based Signature Scheme based on hardware parameters of the sensor node, that is a lightweight solution for authentication problem in WSN. Our scheme achieves this by performing pairing computations in the verification process by the cluster head node and not in the generation process. This is easily computable by the cluster head node which has comparatively lesser constraints on the resources utilized. The significant advantage of our scheme is that the unique structured identity is computed dynamically for every session and is not stored permanently in the sensor node making it impossible for the attacker to extract the identity, thus overcoming the identity based attacks and node capture attacks.

2   Preliminaries

This section presents a brief overview of the definitions and other basic mathematical assumptions followed in this paper.

2.1   Bilinear Pairings

Pairing based cryptography has a number of positive applications like identity-based encryption, identity-based signatures, key agreement and short signatures. The computational capability of sensor nodes are limited, so traditional public key cryptography in which the computation of modular exponentiation is required, cannot be implemented on WSNs [14].

Let G_1 be a cyclic additive group generated by P, whose order is a prime q, and G_2 be a cyclic multiplicative group of the same order q. Let a and b be the elements of Z_q. It is assumed that the discrete logarithm problem (DLP) in both G_1 and G_2 is hard. A bilinear pairing is a map: $\hat{e} : G_1 \times G_1 \to G_2$ with the following properties:

1. **Bilinear:** $\hat{e}(aP, bQ) = \hat{e}(P,Q)^{ab}$.
2. **Non-degenerate:** There exists $P, Q \in G_1$ such that $\hat{e}(P, Q) \neq 1$.
3. **Computable:** For all $P, Q \in G_1$, there is an efficient algorithm to compute $\hat{e}(PQ)$.

2.2   Gap Diffie-Hellman (GDH) Groups

Let G be a cyclic group generated by P, whose order is a prime q. G is assumed to be an additive group, and a, b and c are the elements of Z_q^* [6].

- **Computational Diffie-Hellman Problem (CDHP):** Given (P, aP, bP) where $a, b \in Z_q^*$, compute abP.

- **Decisional Diffie-Hellman Problem (DDHP):** Given (P, aP, bP, cP) where $a, b, c \in Z_q^*$, decide whether $c = ab$ in Z_q^*. (If so, $<P, aP, bP, cP>$ is called a valid Diffie-Hellman tuple).

Definition. A group G is a GDH group, if the DDHP in G can be efficiently computed and there exists no probabilistic algorithm which can solve the CDHP in G with non-negligible probability within polynomial time. If we have an admissible bilinear pairing in G, we can solve the DDHP in G efficiently as follows:

$$(P, aP, bP, cP) \text{ is a valid DH tuple } \iff \hat{e}(aP, bP) = \hat{e}(P, cP)$$

2.3 Identity Based Signatures

An Identity Based Signature scheme (IBS) consists of the following four polynomial-time algorithms:

- **Setup:** Given a security parameter κ, the algorithm generates and outputs the system public parameters along with the master public key mpk, while the corresponding master secret key msk is kept as secret.
- **Extract:** An algorithm which takes as input an identity ID_i as a master secret along with the public parameters of the node if a node and outputs the corresponding private key SK_i.
- **Sign:** An algorithm which takes as inputs a signer's private key SK_i ,identity ID_i and a message m, outputs a signature σ on message m.
- **Verify:** An algorithm which takes as input parameters mpk, a signature σ, a message m and an identity ID_i and outputs *accept* if σ is a valid signature on m for identity ID_i, and outputs *reject* otherwise.

A universally accepted security notion of a signature scheme is existential unforgeability under adaptive chosen message attack (EUF-CMA) in the random oracle model. Under such a model, a forger can query the signing oracle in an adaptive fashion. Its goal is to produce a valid signature on a message that has not been queried to the signing oracle. This model is described as a game EUF-CMA played between a challenger C and a forger $\mathcal{F}$ [13].

2.4 Fingerprint

In a wireless sensor network, every sensor node is loaded with the executable binary code of the application. The beginning part of an application's executable binary code is stored in the sensor node's memory and is termed as the *fingerprint* of a sensor node [17]. This *fingerprint* is lifetime secure against many attacks [16]. Any attacker trying to read the memory contents has to load a malicious code for the same. This code would be forcefully loaded into those memory locations occupied by the fingerprint thus overwriting and hence it is practically unforgeable by an outsider. The hardware in the sensor node ensures that the loaded malicious code gets executed. Hence, the attacker will not able be to read the *fingerprint*, thereby securing the same. For certain applications like surveillance and health care, the node identity need not be unique and fixed one. To authenticate itself to the cluster head node, the sensor node will generate an identity dynamically for every session [23]. The attacker may be interested in the sensitive data collected by the sensor nodes. If the attacker is able to gain access to the data and alter, then data integrity could be violated. Hence it becomes important to authenticate the node and thus the data originating from that node. As the *fingerprint* of a sensor node is life time secure enough against many attacks, we utilize it for authentication purpose. In the proposed scheme, all the nodes in the network including the cluster head node are preloaded with the application before deployment. The length of the fingerprint is fixed for a certain kind of node. This size is decided by the length of the binary code of a minimum-sized malicious program. The *fingerprint* is partitioned into n non-overlapping key elements with a specific length, each of which has a unique identifier S_i where $i \in \{0, 1, \ldots, n-1\}$. Here, the minimum size of the application program is assumed to be 1600 bytes with the length of a key element set to 64 bits [16]. Thus the fingerprint is partitioned into 200 non-overlapping key elements.

3   Proposed Work

In this paper, we propose a hardware based scheme that can be used to augment the security of wireless sensor networks. The main contribution of this work is an authentication scheme to provide security between the sensor node and the cluster node. Rather than relying solely on higher-layer cryptographic mechanisms, wireless sensor devices can authenticate themselves based upon an identity and hence, a physical parameter of a node that can't be compromised is chosen.

In this scheme, a small fraction of the memory where the binary executable of the application program is stored is used as the *fingerprint* of the sensor node. This unforgeable *fingerprint* is used for generating the node identity and digital signature for a session. In addition, the scheme scales properly, being able to manage networks with an arbitrarily large number of nodes.

4   Construction

This is a PKI based signature scheme in the random oracle model under the GDH assumption and will be used by the PKG to generate the private key for the users of the proposed identity based system. The identity of the node is not preloaded in the individual nodes before deployment instead the identity is based on the application code and generated dynamically.

The protocol is modeled as a collection of different nodes $N_1, N_2..N_n$ with identities $ID_1, ID_2 \ldots ID_n$. The node identities are generated dynamically for each session and is not stored in the node permanently, which would overcome the node capture attacks. The identity generation algorithm produces the identity of the node dynamically from the set of valid 64-bit keys for every session. The signature scheme consists of the following four algorithms: **Setup**, **Extract**, **Sign**, and **Verify**.

- **Setup:** Let G_1, G_2 be cyclic prime order groups of order p, where G_1 is an additive group and G_2 is a multiplicative group. Let $P \in_R G_1$ be the generator of G_1, $\hat{e} : G_1 \times G_1 \rightarrow G_2$ be a bilinear map and $H_1(.)$, $H_2(.,.)$ be two cryptographic hash functions defined by,

$$H_1 : \{0,1\}^{64} \rightarrow G_1 \text{ and } H_2 : \{0,1\}^* \times G_1 \rightarrow G_1$$

The PKG chooses $x \in_R Z_p^*$ as the master secret key (msk) and sets the master public key $mpk = xP$. The system parameters are $(P, mpk, \hat{e}, G_1, G_2, H_1, H_2)$.
- **Extract:** For a node N_A, the identity ID_A is computed as follows: The node N_A chooses a random number i in the range $1 < i < 200$ and sends it to the cluster head node, along with the timestamp TS_i. Meanwhile, the cluster node also chooses a random number j $(j \neq i)$ in the range $1 < j < 200$, adds the timestamp $TC_j = TS_i + 1$ and sends it to the node in a secure channel.

The node N_A as well as the cluster head node sets the node's identity as $ID_A = S_i \oplus S_j$ where S_i and S_j are the key elements from the *fingerprint* corresponding to i and j values. These node identities are generated dynamically for every session and the freshness is maintained. The public key PK_A and the private key SK_A are computed as

$$PK_A = H_1(ID_A) \text{ and } SK_A = x.H_1(ID_A).$$

SK_A is the private key of the node N_A corresponding to the public key PK_A. The computed private key SK_A is sent to the node in a secure communication channel.
- **Sign:** To generate the signature on message m, the node N_A using ID_A executes the following algorithm:

    1. The node picks a random value $r \in_R Z_p^*$ and computes $U = r.P$
    2. The node then computes $y_m = H_2(m, U)$ and computes $V = SK_A + r.y_m$
    3. The signature is computed for a message m by the node N_A as
       $\sigma_A = (U, V)$
- **Verify:** On receiving $\sigma_A = (U, V)$, the cluster node checks the validity as

$$\hat{e}(V, P) \stackrel{?}{=} \hat{e}(H_1(ID_A), mpk).\hat{e}(U, y_m) \tag{1}$$

If the above equality holds, then the signature is "valid" for a message m from a node with identity ID_A else "reject".

Correctness: The correctness of the scheme is proved as follows:

$$\hat{e}(V, P) = \hat{e}(SK_A + r.y_m, P)$$
$$= \hat{e}(SK_A, P).\hat{e}(r.y_m, P)$$
$$= \hat{e}(H_1(ID_A), mpk).\hat{e}(U, y_m).$$

5   Security and Performance Analysis

In this section, the security analysis against existential forgery and the resiliency of the scheme against certain identity based attacks are discussed. The performance analysis of the scheme in terms of the computational cost of signature generation, verification are mentioned in this section. Also the results are compared with that of the schemes by Barreto et al. [2], Tso et al. [22] and Narayan and Parampalli [15].

5.1   Security Analysis

The security for the ID-based signature scheme is proved against existential forgery on adaptively chosen message attacks in the random oracle model. In this model, an adversary wins the game if he outputs a valid pair of a message and a signature, where he is allowed to ask the signing oracle to sign any message except the actual message. The scheme is secure against existential forgery on adaptively chosen message and ID based attacks if no polynomial time algorithm $\mathcal{A}$ has a non-negligible advantage against a challenger $\mathcal{C}$ in the following game:

In this model, the challenger $\mathcal{C}$ runs the key generation algorithm to generate a public/private key pair (PK, SK), SK is kept secret while PK is given to the forger $\mathcal{F}$. Further, $\mathcal{F}$ performs a series of oracle queries in an adaptive fashion. The following queries are allowed:

- Hash query: $\mathcal{F}$ submits a string and obtains its corresponding hash value.
- Sign query: $\mathcal{F}$ submits a message (m, ID) to the challenger $\mathcal{C}$ and obtains a signature σ on message m using the private key SK.
- Extract query: $\mathcal{F}$ submits an identity ID to the challenger $\mathcal{C}$ and obtains the private key SK corresponding to the identity ID.

At the end of the game, $\mathcal{F}$ outputs a message and signature pair. $\mathcal{F}$ wins the game, if the output is a valid message-signature pair (m^*, σ^*) with the restriction that m^* has never been asked to the signing oracle. The above game describes a security model for signature unforgeability. $\mathcal{F}$'s advantage is defined to be $Adv(F) = \Pr[\mathcal{F}$ wins the game EUF-CMA$]$. The following theorem shows that the scheme is secure following GDH assumption on the groups (G_1, G_2).

Theorem 1. *Let $\mathcal{A}$ be an EUF-CMA adversary who tries to forge the signature with security parameters κ, in the random oracle model. If $\mathcal{A}$ can break the scheme in time t_1 with success probability ϵ_1, then the CDHP can be solved in time t_2 with success probability ϵ_2 such that $\epsilon_2 \geq \epsilon_1 \left[1 - \frac{1}{q_E} \right] \cdot \left[\frac{1}{q_{H_2}(q_{H_1} - q_E)} \right]$ and $t_2 \leq t_1 + \mathcal{O}(q_E + q_S + q_{H_1} + q_{H_2})$, where $q_E, q_S, q_{H_i} (i = 1, 2)$ are the total number of queries to Extract, Signing and Hash Oracles respectively.*

Proof. Consider $\mathcal{F}$ to be a forger who is assumed to break the signature scheme. If $\mathcal{F}$ can forge the signature scheme, then the challenger $\mathcal{C}$ solves CDHP on (G_1, G_2) with probability at least ϵ_1.

For this, assume P be the generator of G_1 and $(P, aP, bP) \in G_1^3$ be the CDH problem instance given to $\mathcal{C}$. The goal of $\mathcal{C}$ is to find $abP \in G_1$. The challenger $\mathcal{C}$ simulates the oracles and interacts with $\mathcal{F}$ as defined in the EUF-CMA game. The game is viewed as given below:

- **Setup:** The challenger $\mathcal{C}$ runs the setup algorithm and generates the public parameters and $msk.\mathcal{C}$ starts interaction with $\mathcal{F}$ by providing a common string (P, G_1, G_2) where $P \in G_1$, and the public key $P_1 = aP \in G_1$. $\mathcal{C}$ also chooses $0 < \gamma \leq q_H$ randomly and sets the γth unique identity queried to the H_1 hash oracle as the target identity, where q_H be the total number of queries generated. Without loss of generality, we assume ID_γ to be the target identity.
- **Training:** $\mathcal{C}$ interacts with $\mathcal{F}$ using Hash Oracles H_1, H_2, an Extract Oracle and a Signing Oracle as follows:
  - **Hash Oracle H_1:** $\mathcal{C}$ maintains the list L_{H_1}, consisting of tuples of the form $<ID_i, PK_i, r_i>$. When $\mathcal{F}$ queries the oracle H_1 with input ID_i, it responds to $\mathcal{F}$'s queries in the following way:
    * If the tuple $<ID_i, H_1(ID_i), r_i>$ is already available in the L_{H_1} list, retrieve and return $PK_i = H_1(ID_i)$.
    * Otherwise, if $i \neq \gamma$, choose $r_i \in_R Z_p{}^*$ and set $PK_i = r_iP \in G_1$. Store the tuple $<ID_i, PK_i, r_i>$ in L_{H_1} and return PK_i.
    * else if $i = \gamma$, set $PK_i = bP$. Store the tuple $<ID_i, PK_i, r_i>$ in L_{H_1} and return PK_i to $\mathcal{F}$.
  - **Hash Oracle H_2:** The input to this oracle is a pair $<m_i, U_i>$. To respond to the queries by $\mathcal{F}$, challenger $\mathcal{C}$ maintains the list L_{H_2}, having tuples of the form $<m_i, U_i, y_{m_i}, s_i>$, where $y_{m_i} = H_2(m_i, U_i)$. If the tuple is already in the list L_{H_2}, retrieve and return corresponding y_{m_i} to $\mathcal{F}$
  - else, it picks $s_i \in_R Z_p{}^*$, sets $y_{m_i} = s_iP$ and stores $<m_i, U_i, y_{m_i}, s_m>$ in L_{H_2} list and return y_{m_i} to $\mathcal{F}$
- **Extract Oracle:** To respond a query, $\mathcal{C}$ maintains a list L_E, consisting of tuples of the form $<ID_i, SK_i>$. When $\mathcal{F}$ makes a query with ID_i as input, $\mathcal{C}$ checks whether $i = \gamma$, if so, the process is aborted. Otherwise, $\mathcal{C}$ performs the following:
  - If $i \neq \gamma$, then $\mathcal{C}$ searches the list L_E for a matching ID_i. If it exists, $\mathcal{C}$ retrieves SK_i and returns. Otherwise $\mathcal{C}$ searches L_{H_1} and retrieves (PK_i, r_i) and computes the following:
    * Sets $SK_i = r_i mpk = r_ixP = xr_iP$
    * Stores the tuple $<ID_i, PK_i, r_i>$ in L_{H_1} list, the tuple $<ID_i, SK_i>$ in L_E list and return SK_i as the private key to $\mathcal{F}$

Without loss of generality, we assume that any identity is queried only once to this oracle.
- **Signing Oracle:** Let (m_i, ID_i) be the message and identity pair for which $\mathcal{F}$ requests the signature. In response, $\mathcal{C}$ performs the following:
  - If $i \neq \gamma$, then $\mathcal{C}$ runs the sign algorithm as it knows the private key SK_i corresponding to ID_i and returns (U_i, V_i) as the signature on m_i.

- otherwise, if $i = \gamma$, then C performs the following:
  * Queries H_1 oracle with input ID_i and retrieves the entry corresponding to ID_i from L_{H_1}
  * Chooses $r_1 \in_R Z_p$ and set $U_i = r_1(PK_i)$
  * Chooses $r_2 \in_R Z_p$ and set the hash value $H_{m_i} = r_2P - r_1^{-1}P_1$ and store the tuple $<m_i, U_i, y_{m_i}, *>$ in L_{H_2} list
  * Set $V_i = r_1 r_2 PK_i$
- Returns (U_i, V_i) as the signature on the message m_i

Forgery: Eventually, after getting enough training, $\mathcal{F}$ produces a forgery $< m^*, \sigma^* = (U^*, V^*)>$ for ID_i. C aborts if any of the following is true:

- if $i \neq \gamma$ (i.e., ID_i is not the target identity set by the challenger).
- The last field of the tuple corresponding to $<m_i, U_i, y_{m_i}, *>$ in list L_{H_2} is not *
- σ^* corresponding to m^* is invalid

Otherwise, C does the following:

- Retrieves $H_1(ID_\gamma)$ corresponding to ID_γ from the list L_{H_1}.
- Retrieves y_{m_γ} corresponding to $<m^*, U^*>$ from the list L_{H_2}
- Computes $V^* - s_\gamma U^* = SK_\gamma = abP$

It can be proved as follows:

$$
\begin{aligned}
V^* &= SK_\gamma + r_\gamma H_2(m^*, U^*) \\
SK_\gamma &= V^* - r_\gamma H_2(m^*, U^*) \\
&= V^* - r_\gamma y_{m_\gamma} \\
&= V^* - r_\gamma s_\gamma P \\
&= V^* - s_\gamma r_\gamma P \\
&= V^* - s_\gamma U^*
\end{aligned} \tag{2}
$$

The value SK_γ can be equated to

$$
SK_\gamma = bP_1 = baP = abP \tag{3}
$$

Equations (2) and (3) $\iff$ $V^* - s_\gamma U^* = SK_\gamma = abP$. This completes the description of the game between C and $\mathcal{F}$. Now, we show how C solves the CDHP instance (P, aP, bP) with probability at least ϵ_1. For showing this we have to analyze the probability related to the following events:

- E1: C does not abort as a result of Extract query
- E2: $\mathcal{F}$ generates a valid message - signature forgery (m^*, σ^*) for $ID_i = ID_\gamma$, where $\sigma^* = (U^*, V^*)$
- E3: (m^*, U^*, V^*) such that the last field of the tuple corresponding to (m^*, U^*) in the list L_{H_2} is *

Let q_{H_1}, q_E denote the number of queries made to Hash oracle H_1 and Extract oracle respectively. The probabilities of the above events are discussed below:

- Probability of $\mathcal{C}$ aborting during one extract query is $\frac{1}{q_E}$. Thus the probability that $\mathcal{C}$ does not abort in any of the extract queries is $\left[1 - \frac{1}{q_E}\right]$.
- There are totally $q_{H_1} - q_E$ identities eligible for being a valid ID_i and thus $ID_i = ID_\gamma$ happens with probability $\left[\frac{1}{q_{H_1} - q_E}\right]$.
- Assuming E_2 has happened, the probability that $<m_i^*, U^*, y_{m_\gamma}, *> \in L_{H_2}$ is $\left[\frac{1}{q_{H_2}(q_{H_1} - q_E)}\right]$

Now, the probability of Challenger $\mathcal{C}$ solving the CDHP is

$$\epsilon_2 \geq \epsilon_1 \cdot \left[1 - \frac{1}{q_E}\right] \cdot \left[\frac{1}{q_{H_2}(q_{H_1} - q_E)}\right] \text{ and } t_2 \leq t_1 + \mathcal{O}(q_E + q_S + q_{H_1} + q_{H_2}).$$

5.2 Quantitative Performance Analysis

In this section, we evaluate the performance of our scheme in terms of communication overhead. We also give a quantitative analysis of our scheme compared to previous ID-based Signature Schemes - Barreto et al. [2], Tso et al. [22] and Narayan and Parampalli's [15] ID-based signature scheme in terms of the signature size and the computation cost for the signing and verification process. The detailed comparison is given in Table 1. From Table 1, it can be inferred that Narayan's scheme may not be suitable for lightweight devices, such as wireless sensors. Both Barreto's and Tso's schemes require exponentiation operation in the signing stage, which is considered fairly heavy. Due to resource constraints, lightweight devices may not be able to execute such operations. On the other hand, the proposed scheme consists of such heavy operations only in the verification stage. This is easily computable by the cluster head node, which has comparatively lesser constraints on the resources utilized.

Table 1. Efficiency comparison of identity based signature schemes

Scheme	Sign	Verify
Barreto et al. [2]	1SM + 1E + 1H +1A	1P + 1SM + 1E + 1M + 2H + 1A
Tso et al. [22]	1SM + 1E + 3H	1P + 1SM + 1E + 1M + 4H + 1A
Narayan and Parampalli [15]	5E + 1M + 2H	3P + 1E + 1M + 2H
Shim et al. [21]	1E + 3H + 1SM	1P + 1E + 1M+ 3H + 1SR
Proposed Scheme	2SM + 1H + 1A	3P + 1M + 2H

H-Hash, A-Addition, SM-Scalar multiplication
M-Multiplication, P-Pairing, E-Exponentiation

5.3   Resiliency Against Specific Attacks

The resiliency of the proposed scheme against certain attacks are discussed below:

Replay Attack: In replay attack, an attacker replays successfully the previously sent authenticated messages. Because of the wireless multi-hop communication, an adversary could perform a successful replay attack by first preventing the reception of messages and then replaying valid messages at a later point in time. This is a problem that all signature schemes have in common. The addition of timestamp along with the authentication information would serve the purpose of overcoming replay attack.

Node Compromise Attack: As the name suggests, an adversary tries to compromise a sensor node to access all data stored on the node. In symmetric key schemes, where a single key or a subset of keys are used by more than one sensor node to create a private key for a node, a compromise of a single node enables an intruder to generate valid signatures of all sensor nodes sharing that key. Our scheme is resilient to node compromise attacks by the way of using fingerprints for authentication. By the way of its basic architecture, even when the node is compromised, the attacker cannot read the fingerprint of the node as it will be overwritten by the malicious program intended to read the contents. The additional security aspect is that the node IDs are dynamically created at the time of signature generation and not stored in the node's memory.

Sybil Attack: In the Sybil attack, a malicious node will replicate as multiple identities, by either fabricating new identities or impersonate existing ones. In the worst case, an attacker may generate an arbitrary number of additional node identities, using only one physical device, providing a disproportionate amount of influence over the network. The proposed scheme is resilient to this attack since the identity is created at the time of message communication with the help of the cluster node and it is not stored in the node. Hence, the attacker may not be able to find the identity of the node.

6   Conclusion

In this paper, we have designed a lightweight identity based digital signature with the fingerprint for authentication in WSNs. From computation point of view, this scheme requires minimal operations both in signing and verifying than the existing identity based approaches. The main advantage of our signature scheme is that it uses the fingerprint to identify nodes thereby providing security against Sybil attack, node compromise attack, and replay attack. Since the nodes are authenticated with a dynamic identity, the authentication process is not delayed and hence the issues related to late authentication are overcome. Since the identities are dynamically created for every session and not stored in the node, the WSN is scalable. More importantly, attacks created by compromised content-dependent signatures are not possible. This scheme has a limitation of being applicable to WSNs having same application program in all the nodes. In future, the work can be extended for a generic WSN.

References

1. Al-Mahmud, A., Akhtar, R.: Secure sensor node authentication in wireless sensor networks. Int. J. Comput. Appl. **46**(4), 10–17 (2012). Full text available
2. Barreto, P.S.L.M., Libert, B., McCullagh, N., Quisquater, J.-J.: Efficient and provably-secure identity-based signatures and signcryption from bilinear maps. In: Roy, B. (ed.) ASIACRYPT 2005. LNCS, vol. 3788, pp. 515–532. Springer, Heidelberg (2005). doi:10.1007/11593447_28
3. Bellare, M., Namprempre, C., Neven, G.: Security proofs for identity-based identification and signature schemes. In: Cachin, C., Camenisch, J.L. (eds.) EUROCRYPT 2004. LNCS, vol. 3027, pp. 268–286. Springer, Heidelberg (2004). doi:10.1007/978-3-540-24676-3_17
4. Benzaid, C., Lounis, K., Al-Nemrat, A., Badache, N., Alazab, M.: Fast authentication in wireless sensor networks. Future Gener. Comput. Syst. **55**(C), 362–375 (2016)
5. Boneh, D., Lynn, B., Shacham, H.: Short signatures from the weil pairing. J. Cryptol. **17**(4), 297–319 (2004)
6. Choon, J.C., Hee Cheon, J.: An identity-based signature from gap Diffie-Hellman groups. In: Desmedt, Y.G. (ed.) PKC 2003. LNCS, vol. 2567, pp. 18–30. Springer, Heidelberg (2003). doi:10.1007/3-540-36288-6_2
7. Chen, C., Shih, T., Tsai, Y., Li, D.: A bilinear pairing-based dynamic key management, authentication for wireless sensor networks. J. Sens. **2015**, 534657:1–534657:14 (2015)
8. Di Pietro, R., Mancini, L.V., Mei, A.: Energy efficient node-to-node authentication and communication confidentiality in wireless sensor networks. Wirel. Netw. **12**(6), 709–721 (2006)
9. Galindo, D., Garcia, F.D.: A Schnorr-like lightweight identity-based signature scheme. In: Preneel, B. (ed.) AFRICACRYPT 2009. LNCS, vol. 5580, pp. 135–148. Springer, Heidelberg (2009). doi:10.1007/978-3-642-02384-2_9
10. Herranz, J.: Deterministic identity-based signatures for partial aggregation. Comput. J. **49**(3), 322–330 (2006)
11. Hess, F.: Efficient identity based signature schemes based on pairings. In: Nyberg, K., Heys, H. (eds.) SAC 2002. LNCS, vol. 2595, pp. 310–324. Springer, Heidelberg (2003). doi:10.1007/3-540-36492-7_20
12. Libert, B., Quisquater, J.-J.: The exact security of an identity based signature and its applications. IACR Cryptology ePrint Archive 2004:102 (2004)
13. Ma, C., Weng, J., Zheng, D.: Fast digital signature schemes as secure as Diffie-Hellman assumptions. IACR Cryptology ePrint Archive 2007:19 (2007)
14. Mishra, M.R., Kar, J., Majhi, B.: One-pass authenticated key establishment protocol on bilinear pairings for wireless sensor networks. In: 2014 International Conference on Privacy and Security in Mobile Systems, PRISMS 2014, Aalborg, Denmark, 11–14 May 2014, pp. 1–7. IEEE (2014)

15. Narayan, S., Parampalli, U.: Efficient identity-based signatures in the standard model. IET Inf. Secur. **2**(4), 108–118 (2008)
16. Niu, X., Tan, C., Wei, C.: eFKM: an enhanced fingerprint-based key management protocol for wireless sensor networks. In: 2nd International Conference on Networking and Distributed Computing, pp. 299–303. IEEE (2011)
17. Niu, X., Zhu, Y., Cui, L., Ni, L.M.: FKM: a fingerprint-based key management protocol for soc-based sensor networks. In: Wireless Communications and Networking Conference, pp. 1–6. IEEE (2009)
18. Paterson, K.G.: Id-based signatures from pairings on elliptic curves. IACR Cryptology ePrint Archive 2002:4 (2002)
19. Sakai, R., Kasahara, M.: Id based cryptosystems with pairing on elliptic curve. IACR Cryptology ePrint Archive 2003:54 (2003)
20. Seshadri, A., Perrig, A., van Doorn, L., Khosla, P.: SWATT: software-based attestation for embedded devices. In: Security and Privacy, pp. 272–282. IEEE (2004)
21. Shim, K., Lee, Y., Park, C.: EIBAS: an efficient identity-based broadcast authentication scheme in wireless sensor networks. Ad Hoc Netw. **11**(1), 182–189 (2013)
22. Tso, R., Gu, C., Okamoto, T., Okamoto, E.: Efficient ID-based digital signatures with message recovery. In: Bao, F., Ling, S., Okamoto, T., Wang, H., Xing, C. (eds.) CANS 2007. LNCS, vol. 4856, pp. 47–59. Springer, Heidelberg (2007). doi:10.1007/978-3-540-76969-9_4
23. Zhang, J., Shankaran, R., Orgun, M.A., Sattar, A., Varadharajan, V.: A dynamic authentication scheme for hierarchical wireless sensor networks. In: Sénac, P., Ott, M., Seneviratne, A. (eds.) MobiQuitous 2010. LNICSSITE, vol. 73, pp. 186–197. Springer, Heidelberg (2012). doi:10.1007/978-3-642-29154-8_16

Trust Based Data Transmission Mechanism in MANET Using sOLSR

S. Rakesh Kumar$^{(\boxtimes)}$ and N. Gayathri

Bannari Amman Institute of Technology, Sathyamangalam, Erode, India
{rakeshkumar, gayathrin}@bitsathy.ac.in

Abstract. A mobile ad hoc network (MANET) is a type of infrastructure-less network with mobile nodes communicating with each other. The distributed administration and dynamic nature of MANET makes it vulnerable to variety of security attacks. So there is a need to secure the data transmission. The Optimized link state routing protocol is an efficient proactive routing protocol which is suitable for such dense and large scale MANET. In this paper, secure-OLSR (sOLSR) is proposed to ensure the secure path data transmission. A new algorithm has been proposed which calculates the trust values for each node. Based on the trust values, the malicious nodes are found and then a path is selected from the available paths based on the maximum path trust. Simulation results using NS-3 simulator shows that the proposed mechanism is very effective than the existing trust based protocols in terms of packet drop ratio, average latency and overhead.

Keywords: MANET · OLSR · sOLSR · Trust management

1 Introduction

MANET is a kind of wireless ad hoc network that is self-configured by mobile devices communicating with all other nodes freely and dynamically. MANETs has a routable networking environment on top of a Link Layer ad hoc network. Manet has security issues out of which Black-hole attack is one. This attack always have an impact on the routing algorithms. OLSR protocol is optimized link state routing protocol for MANET, which uses hello and topology control (TC) messages to discover and then spread link state information throughout the mobile ad hoc network. The protocol provides an efficient multipoint relays (MPR) selecting mechanism. When comparing with reactive routing protocols, OLSR keeps more stable links and provides better performance in terms of bandwidth and traffic overhead. It is particularly suitable for large and dense mobile network.

1.1 Overview of the OLSR Protocol

The OLSR [26] protocol was published in 2003 by Internet Engineering Task Force. The new version, OLSRv2 [7] was published in 2014. OLSR operations include neighborhood discovery, Multipoint Relay (MPR) selection, topology discovery and

S. Subramanian et al. (Eds.): CSI 2016, CCIS 679, pp. 169–180, 2016.
DOI: 10.1007/978-981-10-3274-5_14

route calculation etc. In OLSR-based MANET, nodes repeatedly broadcast HELLO messages to find its 1-hop and 2-hop neighborhoods. In OLSR, a node selects its MPRs which will forward messages of its selectors during flooding process and further broadcasting continues. Information in the routing table is given below

Destination	Next hop	Distance

In MPR selection process, each node identifies a set of its MPR nodes that can broadcast its routing messages. In OLSR, a node finds its MPR set that can reach all its two-hop neighbors. In case there are multiple choices, the minimum set is selected as an MPR set.

In OLSRv2 [7] MPR set is a subset of 1-hop neighbors, but with minimum link metrics. In OLSRv1, MPR set is also a subset of 1-hop neighbors but the element nodes are able to reach all of its 2-hop neighbor's (i.e. the minimum hop counts). To advertise the link to its selectors, an MPR generates Topology control (TC) messages that contain certain link information [17]. These TC messages will be flooded to the entire network. As a result, a node will know the information of the entire topology after receiving certain TC messages.

2 Related Works

Recently several techniques had been contributed for securing OLSR. Li [1] had proposed a trust reasoning model based on fuzzy Petri net to evaluate trust values of mobile nodes. In addition, to avoid compromised nodes, a trust based routing algorithm is proposed which will help to select a path with the maximum trust value among all possible paths.

Abdalla et al. [2] proposed a mechanism in which Path validation message (PVM) and Path validation message back (PVMB) are sent to check the malicious node in a data forwarding path. If a malicious node is detected then Attacker Finder Message (AFM) and Attacker Finder Message back (AFMB) are used to detect its address. Then the corresponding node with that address would be alerted by Attacker Isolation Message (AIM) such that it is isolated from the entire network. Boukerch et al. proposed a mechanism called novel agent-based trust and reputation management scheme (ATRM) in [3]. Using this scheme requires paying close attention to the bandwidth and delay overhead. The major aim of this scheme is to manage trust and reputation locally with minimal overhead.

Jia et al. [4], proposed a trust model, which is multi-path routing protocol and is of reactive nature, named as ad hoc on-demand trusted multi-path (AOTMDV) routing protocol in [5]. This protocol provides a flexible approach to choose the shortest path which will meet the security requirements of packet transmission. The effectiveness of the proposed mechanism in identification of the compromised nodes is evaluated based on various real time experiments. Xia et al. [5], proposed a trust prediction model to calculate the trustworthiness of nodes, which is based on nodes historical behavior as well as future behaviors using extended fuzzy logic rules. Adnane et al. [6], proposed trust specification language and showed how trust based reasoning can allow each node to evaluate the behavior of other nodes. After finding the misbehavior nodes, the proposed solution takes necessary measures to resolve the situation of inconsistency and counter the malicious nodes. Efficient broadcast encryption scheme for key distribution in MANET is chosen Cipher text attack (CCA). In this, no message exchange is required to establish a group key [8]. Novel trust based mechanism for AODV is proposed in [10]. A trust model which evaluates neighbor's direct trust by factors such as mobility and cooperation frequency. To diminish vulnerabilities of OLSR protocol, a new method is presented which is trust based reasoning in the nodes [9, 11]. A knowledge based acquisition and representation approach using the fuzzy reasoning and dynamic adaptive FPNs to solve the problems is proposed in [12, 13]. Security mechanisms which combat the security issues have been proposed in [14]. The system survivability is critically low because of the selfish nodes. To tackle that a trust management approach is proposed in [15]. The comparative study between the behavior modes of the node with the OLSR specification is briefly explained in Trust Reasoning model [23]. The formal specification for trusted neighbor is proposed by Verma in [16]. In [18] a trust-based reactive multipath routing protocol is proposed. It is used to discover multiple loop-free paths as candidates in one route discovery. A composite trust metric is derived from social trust and quality of service (Qos) in [19].

Extension to OLSR protocol is presented in [20, 25]. In this method, security features are added to existing OLSR protocol. Security features are based on authentication checks of information injected into the network, mutual authentication between two nodes. Countermeasures are presented to prevent the attacks against OLSR [21]. Kishore deals with collision attack [22]. Collusion attack is prevented by incorporating an information theoretic trust framework in OLSR. The comparative study between the behavior modes of the node with the olsr specification is briefly explained in Trust Reasoning model [23] If a node does not send TC message for a long period of time it is identified and isolated in [24]. An effective trust evaluation system is proposed [27] by performing extensive simulations. A secured route is found by the source node using the trust levels [28] and then the message is transmitted. A central trust authority is demonstrated by presenting a model for trust based communication in [29].

In this paper, an attack based routing mechanism is proposed to deal with the malicious node and prevent the node from attacking the network. In this mechanism, the trust worthiness of each node is calculated and based on that value; the malicious node in the network is detected. For routing, each node separately has a routing table. Based on the trust worthiness, the nodes update their routing table and select the path which has the maximum trust value.

3   Proposed Work

In this section, an attack based routing mechanism is presented to find the secure path for data transmission and also the energy consumed by each node is calculated. Trust Evaluation algorithm is the one which aids in calculating the trust value for further examination and selection of a trustable path.

3.1 Trust Evaluation Algorithm

In MANET, Black hole means a node in the network which discards the messages without informing the source that the message is not reached by the destination. The proposed algorithm is as follows which helps in efficient trust evaluation.

The initial tokens (S), input incidence from places the transitions (W), output incidence from transition to places (U), and the threshold (H) is taken as the input. The trust value (E^{trust}) and non trust value (E^{-trust}) are recorded as the output.

Algorithm

```
Input:      S (n), W, U, H, n=0
Output: Etrust, E-trust, Euncertain
Step1: R=W . S (n)
Step2: If R T > H THEN M = R T
            Else M = 0
Step3: Sn+1= max {U , M }
Step4: Sn+1= max { Sn , Sn+1 }
Step5: If Sn+1=S (n) go to step6
            Else replace S (n) with Sn+1
Step6: If S`(n) ij + S (n) (i(j+1))< 1
                        Etrust = S (n)
                                       (i(j+1))
                    E-trust =  S (n)ij
          Euncertain =1-S (n)n-S (n) (i(j+1))
          Else  Etrust = S(n) (i(j+1))
                Euncertain = 0
                E-trust = 1-S(n)
                              (i(j+1))


          end if
```

Step1: The input incidence is multiplied with the initial tokens and it is assigned to a matrix variable (R).
Step2: If the transpose of the matrix is less than the threshold the transpose matrix is assigned to another variable (M) otherwise the value is assigned to 0.
Step3: The initial token of the next node is calculated from the maximum value of output incidence and the variable M.
Step4: The initial token of the next node is calculated from the maximum value of current node and the next node value.

Step5: If the initial token of next and current node value is equal then go to step6 otherwise replace the value of current node with the previous node value. Step6: If sum of the value of current and next initial token is less than 1 then the trust value is equal to next initial token value and non trust value is equal to current initial token value and the remaining is uncertain value. In the other case, if the sum is more than one, then the trust value is equal to next initial token value and uncertain is equal to 0 and non trust value is equal to remaining value.

3.2 Secured OLSR

An attack based routing technique sOLSR aims to protect the data from malicious or compromised nodes. This is done by selecting a path which has the maximum path trust value among all the possible paths between any pair of nodes. Before that the path trust value of each node must be calculated. Based on that value and the information from the other nodes, it is decided that the node is malicious or not. If the node has minimum path trust value, then that node is eliminated from the data forwarding route.

3.2.1 Path Trust Value

Each node has a trust value. While forwarding a data from source to destination, it checks all the possible paths. From that it selects the path with the maximum trust value. For calculating the trust value, initialization of the trust factors for every node is done. From the trust factors, trust value is generated using the trust evaluation algorithm. While creating a routing table, a node needs to manage two tables: the trust value based routing table TAB_{PT} and the register table TAB_{REG}. TAB_{PT} records the path with the maximum path trust value, and TAB_{REG} has all possible paths from source to destination. An entry in a routing table is labeled as Entry = (Dest, PT, Next Hop) where Dest is the address of destination node, PT is the trust value, and Next Hop is the address of next hop. OLSR enables a node to know the entire network topology information. Consider the network as shown in Fig. 1.

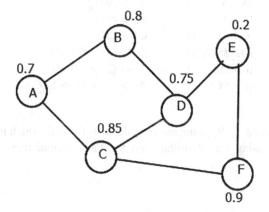

Fig. 1. Network depicting the trust value of every node

Consider the routing table for A. At first TAB_{PT} has its first hop neighbors and TAB_{REG} has the second hop neighbors as in Table 1. From TAB_{REG}, the path with the maximum trust value is selected.

Table 1. Routing algorithm calculation

TAB_{PT}	TAB_{REG}
(A,1,A)	(D,0.8,B)
(B,1,B)	(F,0.85,C)
(C,1,C)	(1,0.3,C)
(D,0.85,C)	(H,0.3,C)

From the TAB_{REG} Table, TAB_{PT} selects the next hop neighbor which is having the highest trust value. And the table TAB_{PT} is updated for each node in the network as in Table 2.

Table 2. Updated table

TAB_{PT}	TAB_{REG}
(A,1,A)	(D,0.8,B)
(B,1,B)	(F,0.85,C)
(C,1,C)	(1,0.3,C)
(D,0.85,C)	(H,0.3,C)

Finally the table will be like this.

Table 3. Final routing table

TAB_{PT}	TAB_{REG}	PATH
(A,1,A)	Ø	A—> A(1)
(B,1,B)	Ø	B—> B(1)
(C,1,C)	Ø	C—> C(1)
(D,0.85,C)	Ø	A—> C—> D(0.85)
(F,0.85,C)	Ø	A—> C—> F(0.85)
(E,0.8,C)	Ø	A—> C—> D—> E(0.8)

Table 3 shows that TAB_{PT} contains only the selected path which has the maximum trust value. Thus using the algorithm and the table formulation, the best path is selected.

4 Flow of the Mechanism

Figure 2 shows the flow of the proposed mechanism. The initialization of number of nodes (say-20 nodes) is done. Then initialization of the trust factors namely load, packet forwarding rate, average forwarding rate and protocol deviation flag of each node is done. By using trust evaluation algorithm, the trust value of each and every node is calculated and malicious node is identified using that value. Routing table is subsequently created. The path with the highest trust value is selected.

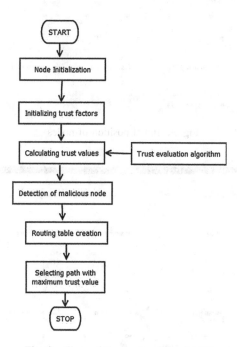

Fig. 2. Flow of the mechanism sOLSR

5 Simulation Result

The proposed algorithm for malicious node detection in MANETs using OLSR protocol is assessed for its performance. The results are simulated using ns3.20 simulator. The assumption is that all the nodes are moving dynamically.

Figure 3 shows the initial position of the nodes, where node 1 is considered as malicious node. Node 4 is assumed to be the source and Node 19 as destination. In Fig. 4 the node 1 which is the malicious node receives messages from the other nodes but does not transfer any message further. Routing table for this network is created and from that the secure path is identified based on the proposed algorithm. It is clearly shown in the routing table in Fig. 5 that node1 does not send any messages to other nodes and hence the particular node is eliminated from the secure path.

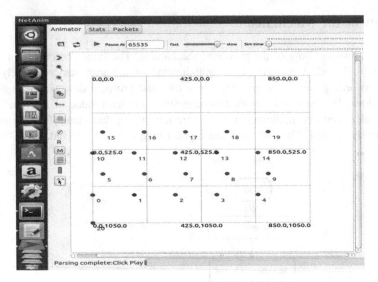

Fig. 3. Initial position of nodes

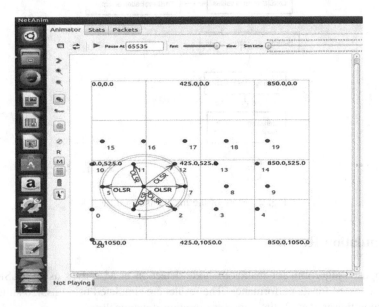

Fig. 4. Malicious node receives message

5.1   Results

The parameters for simulation are listed in the table below:

Table 4 shows that 50 nodes are randomly placed in a 1000 m * 1000 m rectangular area. End-to-end data traffic flow at the rate of 300 kbps is set. The performance of the protocol is evaluated using three metrics namely.

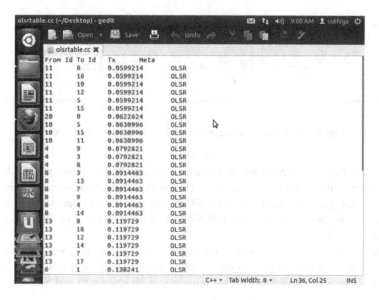

Fig. 5. Routing entry for the malicious node

Table 4. Simulation parameters

Parameter	Value
Topology area	1000 m * 1000 m
Number of nodes	50
Node moving speed	0–10 m/s
Bandwidth	2 Mbps
Data rate	300 Kbps
Number of data traffic flow	3
Simulation time	100 s
Transmission radius	250 m
Number of malicious nodes	0–10

a. Packet Delivery Ratio: It is defined by the ratio of packets that are successfully delivered to a destination when compared to the number of packets that have been sent out by the sender.

b. Average End-to-End Latency: It represents the average time incurred in transmitting data packets from source to destination.

c. Routing packet Overhead: Average number of routing control packet.

The performance of OLSR and sOLSR with varying number of malicious nodes is studied. Figure 6(a) shows that sOLSR performs better than OLSR in terms of packet delivery ratio. The packet delivery ratio of sOLSR is not 100% because of the fact that the malicious nodes may be used as intermediate node to forward the data until trustable routes are established.

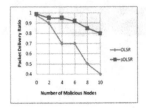 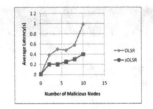 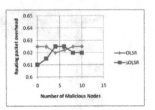

Fig. 6. Performance comparison with varying number of nodes

Figure 6(b) gives the comparison of the average latency for OLSR and sOLSR. The average latency of OLSR is low because only few data packets are delivered through short routes. sOLSR performs better than OLSR in terms of average latency. Figure 6(C) gives the routing overhead values of sOLSR and OLSR. Overhead of OLSR is comparable with sOLSR.

Figure 7 gives the performance measures with respect to speed of the node. Figure 7(a) gives packet delivery ratio which is better with respect to sOLSR in terms of speed of the node. Overhead is also minimum with respect to sOLSR.

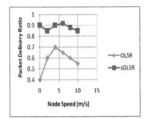 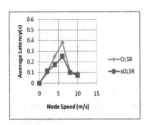 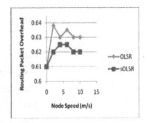

Fig. 7. Performance comparison with varying mobile speed

6   Conclusion

Security of the packet transfer in a network is being addressed in this paper. A secure-OLSR (sOLSR) is proposed to remove the malicious node from the network. A secure path is selected based on the trust value of the node. The proposed algorithm is evaluated against OLSR for its performance in terms of packet delivery ratio, latency and routing overhead. Simulation results show that the performance of sOLSR is far better than OLSR with varying number of malicious nodes as well as varying mobile speed. In future, the sequence number could also be considered for trust value calculation to improve its performance.

References

1. Tan, S., Li, X., Dong, Q.: Trust based routing mechanism for securing OSLR-based MANET. Ad Hoc Netw. **30**, 84–98 (2015)

2. Abdalla, A.M., Saroit, I.A., et al.: Misbehavior nodes detection and isolation for MANETs OLSR protocol. Procedia Comput. Sci. **3**, 115–121 (2011)
3. Boukerch, A., Xu, L., El-Khatib, K.: Trust-based security for wireless adhocand sensor networks. Comput. Commun. **11**, 2413–2427 (2007)
4. Xia, H., Jia, Z., et al.: Impact of trust model on on-demand multi-path routing in mobile ad hoc networks. Comput. Commun. **36**, 1078–1093 (2013)
5. Xia, H., Jia, Z., et al.: Trust prediction and trust-based source routing in mobile ad hoc networks. Ad Hoc Netw. **11**, 2096–2114 (2013)
6. Adnane, A., Bidan, C., de Sousa, R.T.: Trust based security for the OLSR routing protocol. Comput. Commun. **36**, 1159–1171 (2013)
7. Clausen, T., Dearlove, C., Jacquet, P., Jia, Z.: The Optimized Link State Routing Protocol Version 2, IETF RFC7181, pp. 1–115 (2014)
8. Yang, Y.: Broadcast encryption based non-interactive key distribution in MANETs. J. Comput. Syst. Sci. **80**, 533–545 (2014)
9. Obaidat, M.S., Woungang, I., Abdalla et al.: A cryptography based protocol against packet dropping and message tampering attacks on mobile ad hoc networks. Secur. Commun. Netw. **7**, 376–384 (2014)
10. Feng, R., Che, S.: A credible routing based on a novel trust mechanism in ad hoc networks. Int. J. Distrib. Sens. Netw. (2013)
11. Adnane, A., et al.: Autonomic trust reasoning enables misbehavior detection in OLSR. In: Proceedings of the 2008 ACM Symposium on Applied Computing, pp. 2006–2013 (2008)
12. Huang, M., Lin, X., Hou, Z.W.: Modeling method of fuzzy fault Petri nets and its application. J. Central South Univ. (Sci. Technol.) **44**, 208–215 (2013)
13. Liu, H.C., Liu, L.: Knowledge acquisition and representation using fuzzy evidential reasoning and dynamic adaptive fuzzy Petri nets. IEEE Trans. Cybern. **43**, 1059–1072 (2013)
14. Abdelaziz, A.K., Nafaa, M., Salim, G.: Survey of routing attacks and countermeasures in mobile ad hoc networks. In: 15th International Conference on Computer Modelling and Simulation (UKSim), pp. 693–698 (2013)
15. Cho, J.H., Chen, I.R.: On the tradeoff between altruism and selfishness in MANET trust management. Ad Hoc Netw. **11**, 2217–2234 (2013)
16. Verma, S., Gujral, M.: Formal specification of trusted neighbor information base of OLSR routing protocol of adhoc network using Z language. In: Krishna, P.V., Babu, M.R., Ariwa, E. (eds.) Global Trends in Computing and Communication Systems, pp. 560–570. Springer, Heidelberg (2012)
17. Robert, J.M., Otrok, H., Chriqi, A.: RBC-OLSR: reputation-based clustering OLSR protocol for wireless ad hoc networks. Comput. Commun. **35**, 487–499 (2012)
18. Li, X., Jia, Z., Zhang, P., et al.: Trust-based on-demand multipath routing in mobile ad hoc networks. IET Inf. Secur. **4**, 212–232 (2010)
19. Cho, J.H., Swami, A., Chan, I.R.: A survey on trust management for mobile ad hoc networks. IEEE Commun. Surv. Tutor. **9**(4), 562–583 (2011)
20. Rana, S., Kapil, A.: Defending against node misbehavior to discover secure route in OLSR. Commun. Comput. Inf. Sci. **70**, 430–436 (2010)
21. Adnane, A., Bidan, C., de Sousa, R.T.: Trust-based countermeasures for securing OLSR protocol. In: International Conference on Computational Science and Engineering, vol. 2, pp. 745–752 (2009)
22. Babu, M.N.K., Franklin, A.A., Murthy, C.S.R.: On the prevention of collusion attack in OLSR-based mobile ad hoc networks. In: 16th IEEE International Conference on Networks (ICON), pp. 1–6 (2008)

23. Adnane, A., de Sousa Jr., R., Bidan, C., et al.: Integrating trust reasonings into node behavior in olsr. In: Proceedings of the 3rd ACM Workshop on QoS and Security for Wireless and Mobile Networks, pp. 152–155 (2007)

24. Kannhavong, B., Nakayama, H., et al.: Analysis of the node isolation attack against OLSR-based mobile ad hoc networks. In: International Symposium on Computer Networks, pp. 30–35 (2006)

25. Hafslund, A., Tønnesen, A.: Secure extension to the OLSR protocol. In: Proceedings of the OLSR Interop and Workshop (2004)

26. Clausen, T., Jacquet, P.: Optimized link state routing protocol (OLSR). In: IETF RFC3626, pp. 1–75 (2003)

27. Sun, Y.L., Han, Z., Yu, W.: Attacks on trust valuation in distributed networks, pp. 1461–1466, March 2006

28. Liu, Z., Joy, A.W., Thompson, R.A.: A dynamic trust model for mobile ad hoc networks, pp. 80–85 (2004)

29. Pirzada, A.A., McDonald, C.: Trust establishment in pure ad hoc network. Wireless Pers. Commun. 37(1), 305–319 (2006)

A Novel Combined Forecasting Technique for Efficient Virtual Machine Migration in Cloud Environment

Getzi Jeba Leelipushpam Paulraj[(✉)], Sharmila John Francis,
and Immanuel John Raja Jebadurai

Karunya University, Coimbatore, India
getzi@karunya.edu, sharmilaanand2003@yahoo.co.in,
immanueljohnraja@gmail.com

Abstract. Live virtual machine (VM) migration relocates running virtual machine from source physical server to the destination physical server without compromising the availability of service to the users. Live VM Migration guarantees energy saving, fault tolerance and uninterrupted server maintenance for the cloud datacenter. The workload handled by the cloud datacenters are unpredictable in nature. Hence, the migration needs intense planning. Resource starvation occurs due to dynamic nature of workload handled by cloud datacenter. The objective of this paper is to predict the resource requirement of the virtual machines running various workloads and to appropriately place them during migration. The resource requirement of the running virtual machines are predicted using combined forecast technique. The combined forecasting technique improves the forecasting accuracy. Every host machine suitably migrates based on the current and forecasted utilization. The proposed algorithm has been validated using set of simulations conducted on Google Datacenter Traces. The results show that the proposed methodology improves the forecasting accuracy.

Keywords: Neural networks · Combined forecasting · Cloud datacenter · Virtual machines migration

1 Introduction

Virtualization has been a meritorious implementation to cloud data centers in managing its resources efficiently. Live Virtual machine Migration is a technique that migrate the entire OS and its associated application from one physical machine to another. The Virtual machines are migrated lively without disrupting the application running on it. The benefits of virtual machine migration include conservation of physical server energy, load balancing among the physical servers and failure tolerance in case of sudden failure [6]. However the application running in cloud datacenters are dynamic in nature. Improper migration without considering the dynamic nature of workload increases the number of migration. This degrades the performance of cloud datacenter [4]. The dynamic nature of the workload is predicted using the forecasting techniques. The prediction is used to migrate the virtual machine from source physical server to the destination physical server to reduce the migration overhead.

© Springer Nature Singapore Pte Ltd. 2016
S. Subramanian et al. (Eds.): CSI 2016, CCIS 679, pp. 181–190, 2016.
DOI: 10.1007/978-981-10-3274-5_15

This paper focuses on predicting the resources of physical servers using combined forecasting techniques. Many classical prediction techniques have been proposed in the literature for forecasting the dynamic nature of workload. However to improve the accuracy an attempt has been made to propose a novel methodology to combine the classical techniques for forecasting. Adequate simulations and analysis have been carried out to investigate the performance of the proposed methodology. The results show that proposed combined forecasting methodology improves the accuracy.

The remainder of the paper is organized as follows. Section 2 presents the existing methods for live virtual machine migration. Details of the proposed combined forecasting techniques is given in Sect. 3. The experimentation and result analysis are provided in Sect. 4. Section 5 gives the conclusion and future scope.

2 Related Works

Live virtual machine migration are classified into Pre-copy and Post copy migration techniques. In Pre-copy [2] technique the destination physical server is selected and the resources are reserved. The memory pages are iteratively copied from source to the destination physical server until the stop and copy phase. During the stop and copy the virtual machine is stopped, the remaining memory pages and the CPU state are copied from source physical server to the destination physical server. The virtual machine is then resumed at the destination physical server. In Post-copy approach the minimal CPU state information and memory pages are transferred to the destination physical server. Then the memory pages required by the running process are fetched from source physical server to the destination physical server through page faults. Wood et al. [5] proposed a migration technique by considering the load of the physical servers. The hotspots are monitored and detected and the overloaded virtual machines are migrated to the under loaded physical servers. Monitoring and detecting such sudden spikes is challenging in a distributed environment. One way of proactively handling the sudden spike is to predict the workload handled by cloud clusters. Combined forecasting technique proved to be the most accurate technique for workload prediction [1, 10, 11]. Using the combined forecasting output suitable decision is made to migrate the virtual machine to the suitable destination. This could prevent further migration and could reduce migration overhead.

3 Proposed Methodology – Combined Forecasting Module

Combined forecast method combines several forecasts made through classical forecasting models [9]. The resource utilization of all the physical machines is forecasted using combined forecasts and the virtual machines are placed suitably to manage sudden spikes in cloud environment. The classical forecasting methods combined in this paper are Exponential smoothing average, Holt winter's method, Autoregressive model and ARIMA model. The block diagram for the combined forecasting technique is shown in Fig. 1.

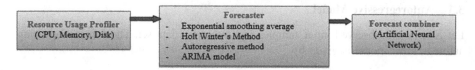

Fig. 1. Block diagram for combined forecasting techniques

CPU, Memory and Disk usage details are captured by resource usage profiler. The output is given to the forecaster and then to the forecast combiner for accurate forecasting.

3.1 Exponential Smoothing Average

The past observations are weighted in exponentially decreasing order to forecast the future value. Let Y_t be the observed samples and S_t be the predicted sample. If the value of α is small, more weightage is given to the sample in more distant past. If the value of α is large, more weightage is given to the recent observations.

$$S_t = \alpha Y_{t-1} + (1 - \alpha)S_{t-1} \quad 0 < \alpha \le 1 \tag{1}$$

In the proposed methodology, the value of α is selected to be 0.9. The pseudocode for the Exponential Smoothing Average Forecasting is given in Pseudocode 1.

```
Exp_R_{t+1} = exp_R ( R_t)      where R = CPU, Memory, Disk
function f = exp_R(R_util)
    α = 0.9;
    f(1)= 0;
    for i = 1 : N
            f(i+1)= (R_util(i) * α) + ((1- α) * f(i));
end
```

Pseudocode 1. Exponential Smoothing Average Forecasting

3.2 Holt Winter's Method

Holt winter's method considers the seasonal changes in the data set. Seasonality occurs when a specific pattern repeats over a period of L in the time series data. It also captures the trend fashioned in the time series data. Suppose Y_t be the observed sample that exhibits seasonal behavior in every period P. Let s (t) be the smoothing value. Let T_t denotes the linear trend that was hidden inside the seasonal changes. C_t be the seasonal correction factor. The pseudocode to calculate the forecasted output F (t + m) using Holt Winter method is given in Pseudocode 2.

```
S_0 = x_0
S_t = α (Y_t − c_{t-P}) + (1-α) (S_{t-1} + T_{t-1})
T_t = β (s_t − s_{t-1}) + (1-β) Tt_1
C_t = γ Y_t / s_t + (1-γ) c_{t-P}
                    F_{t+m} = (s_t + m T_t) C_{t-P+1+ (m-1) mod P}
```
Pseudocode 2. Holt winter's method

3.3 Autoregressive Model

Let X(t) be the observed input samples in time series data set. The forecasting using autoregressive model is given by

$$X(t) = c + \sum_{i=1}^{i=p} \Phi i * x(t+i)$$ (2)

C is the constant. The value of Φ is calculated using MATLAB function aryule for the regression of three past samples. The Pseudocode for Autoregressive model is given in Pseudocode 3.

```
ar_R_{t+1} = ar_R ( R_t )          where R = CPU, Memory, Disk
function f = ar_R(R_util)
Mdl [a,b] = aryule(R_util,2);
for  I = 1:N
        R_util(t+1) = a + b * R_util(t) + c* R_util(t-1)
end
```

Pseudocode 3. Autoregressive Model

3.4 Autoregressive Integrated Moving Average (ARIMA) Model

The Autoregressive integrated moving average model aims at removing the non-stationary behavior from the time series data. The regression part of ARIMA model indicates that the future values strongly depend on its lagged values and the moving average part takes the past error to predict the future values. The forecasting using ARIMA model is given by

$$\hat{y}_t = \mu + \phi_1 y_{t-1} + \ldots + \phi_p y_{t-p} - \theta_1 e_{t-1} - \ldots - \theta_q e_{t-q}$$ (3)

Here y_t indicate the observed time series data and y_{t-1} is the lagged form of the observed samples. Pseudocode 4 gives the Autoregressive Integrated Moving Average Model

```
arima_R_{t+1} = arima_R ( R_t )  where R = CPU, Memory, Disk
function f = arima_R(R_util)
for i = 1 : N
Mdl = arima (2,1,1);
Estmdl = estimate (mdl, R_util);
R_util(t+1) = forecast(estmdl,R_util);
End
```

Pseudocode 4. The Autoregressive Integrated Moving Average Model

3.5 Neural Networks

The neural network back propagation method is used to combine all the output of the classical forecasting techniques [6]. Four nodes per forecasting technique have been selected in the input layer. To combine the output one node has been selected in the hidden and output layer. The weights in the input layer is given in Eqs. (4) to (7).

$$w\_input(1) = 1- mse\_esa_t \qquad (4)$$

$$w\_input(2) = 1- mse\_hw_t \qquad (5)$$

$$w\_input(3) = 1- mse\_ar_t \qquad (6)$$

$$w\_input(4) = 1- mse\_arima_t \qquad (7)$$

By subtracting the weight with one, the method with more error is given less weightage. The output of the input layer are given in Eqs. (8) to (11),

$$Output\_input_t = linear\ (forecast\_esa_t) \qquad (8)$$

$$Output\_input_t = linear\ (forecast\_hw_t) \qquad (9)$$

$$Output\_input_t = linear\ (forecast\_ar_t) \qquad (10)$$

$$Output\_input_t = linear\ (forecast\_arima_t) \qquad (11)$$

The Linear method is applied in the input layer as the forecasted output with high error is given less weightage when combined in the hidden and output layer. Sigmoidal function is applied in the hidden layer as given in Eq. (12).

$$output\_hidden_t = 1/(1-exp(input\_hidden_t)) \qquad (12)$$

The output to the hidden layer is multiplied by weights and applied as the input to the output layer to find the predicted value at time t as given in Eq. (13).

$$output_t = 1/(1-exp(input\_output_t)) \qquad (13)$$

The error is calculated as predicted value subtracted from the actual value [12]. The weights in the hidden and the input layers are updated as δW and δV. The iteration continues till the error value is minimum.

4 Proposed Methodology – Migration Module

Using combined forecasting technique the future CPU and memory requirement of individual server and virtual machines are calculated. The servers are classified into four different sets in the given forecasting window W. The detail of the classification set is given in Table 1.

The Source virtual machine request the migration controller for migration. The source virtual machine holds the current resource requirement $R_c = \{R_{cpu}, R_{memory}\}$, where R_{cpu} stands for the number of MIPS required to complete the task running in the virtual machine and R_{memory} stands for Memory space required for completing the task

Table 1. Classification of server farm

Server farm ID	Notation	Resource requirement in forecasting window W
S_0	$\{H_c, H_m\}$	Higher CPU and Memory requirement
S_1	$\{Hc, Lm\}$	Higher CPU and Lower memory requirement
S_2	$\{Lc, Hm\}$	Lower CPU and Higher memory requirement
S_3	$\{L_c, Lm\}$	Lower CPU and memory requirement

running in the migrating virtual machine. Based on the resource requirement R_c servers S_n are selected from the server farm. The forecasted resource requirement R_f is the considered for target selection. The virtual machine requesting for migration falls in any one of the four categories viz., $\{L_c, Lm\}$, $\{Lc, Hm\}$, $\{Hc, Lm\}$, $\{H_c, H_m\}$. Based on the criteria server S_m is selected. The source virtual machine is placed in the target server based on the pre-copy based migration algorithm. The Flow diagram for combined forecasting based pre-copy migration algorithm is depicted in Fig. 2. The Pre-Copy Migration algorithm is explained in the Pseudocode 5.

```
Initialize Datacenter D
Server List S € Datacenter D
VMn - Number of virtual machine running in each server
  For VM = 1 to VMList
      Get Rc = {Rcpu, Rmemory}
      Combined forecast (Rc) = Rf        where Rf = {ForecastCPU, Forecastmemory}
  End
For server = 1 to serverList
      Get Rc = Σ₁^VMn Rc
      Get Rf = Σ₁^VMn Rf
End
For Server S € Server List
    Set Forecasting window W
    Classify (ServerList, Rf, W)
            If (Rcpu > ForecastCPU & Rmemory > Forecastmemory ) then
            S € S0 = {Hc, Hm}
            ElseIf (Rcpu > ForecastCPU & Rmemory < Forecastmemory ) then
            S € S1 = {Hc, Lm}
            ElseIf (Rcpu < ForecastCPU & Rmemory > Forecastmemory ) then
            S € S2 = {Lc, Hm}
            Else
            S € S3 = {Lc, Lm}
End
Migration (VM € VmList )
            If VM € S0 then
            Migrate (VM -> S3 = {Lc, Lm}
            Elseif VM € S1 then
            Migrate (VM -> S2 = {Lc, Hm}
            Elseif VM € S2 then
            Migrate (VM -> S1 = {Hc, Lm}
            Else VM € S3 then
            Migrate (VM -> S0 = {Hc, Hm}
            End
```

Pseudocode 5. Virtual machine Placement algorithm

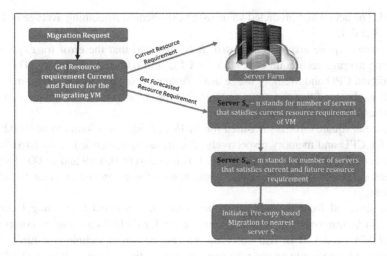

Fig. 2. Combined Forecasting based Pre-copy migration flow diagram

5 Performance Analysis

5.1 Combined Forecasting Technique

The proposed methodology of combined forecasting techniques is simulated in MATLAB 2014. Google cluster 2011-2 has been used as the trace data. These are data collected from 12.5 K machines over a period of 29 days. The trace includes normalized resource utilization details of CPU (core count), memory (bytes) and disk space (bytes). The CPU and Memory Data for 50 servers are segregated from the google data trace. The classical forecasting techniques have been implemented. Analysis were carried out to investigate the performance of the classical methods in forecasting the future CPU and Memory Load. The actual CPU and Memory rate against the predicted rate were

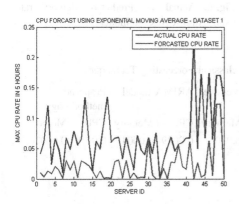

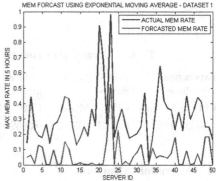

Fig. 3. Actual vs. Predicted CPU rate using Exponential smoothing average

Fig. 4. Actual vs. Predicted Memory rate using Exponential smoothing average

obtained. The actual and predicted value using exponential smoothing average is shown in Figs. 3 and 4.

The mean square error is calculated. It is observed that the error for exponential smoothing average technique is 0.9455 for CPU and 0.0621 for memory. The actual and predicted CPU and Memory usage using Auto regressive model is shown in Figs. 5 and 6. It is observed that the mean square error for autoregressive technique is 0.0069 for CPU and 0.0005 for memory.

The mean square error is measured for ARIMA model. It is found to be 0.0029 and 0.0004 for CPU and memory respectively. The mean square error is calculated for the proposed combined forecasting technique. It is found to be 0.0019 and 0.0002 for CPU and memory respectively. The Table 2 summaries the mean square error for all the techniques.

It is observed from the table that the proposed combined forecasting technique exhibits 34 % improvement in mean square error for CPU forecasting in comparison with ARIMA model. It is also observed that the proposed technique exhibits 50 % improvement in mean square error for memory forecasting in comparison with ARIMA model.

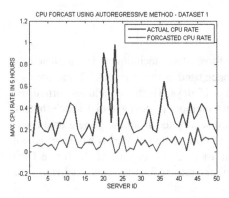 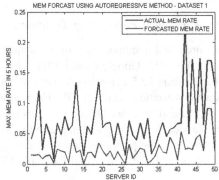

Fig. 5. Actual vs. Predicted CPU rate using Auto regressive model

Fig. 6. Actual vs. Predicted Memory rate using Auto regressive model

Table 2. Mean Square error for classical Forecasting Techniques

Exponential smoothing		Holt Winter's method		Auto Regressive method		ARIMA model		Proposed methodology	
CPU	Memory	CPU	Memory	CPU	Memory	CPU	Memory	CPU	Memory
0.9455	0.0621	0.0122	0.0122	0.0069	0.0005	0.0029	0.0004	0.0019	0.0002

5.2 Migration Algorithm

The virtual machine migration is simulated using Cloudsim Toolkit [3]. The experimental setup has one datacenter and 50 servers. The Google data Trace is used to

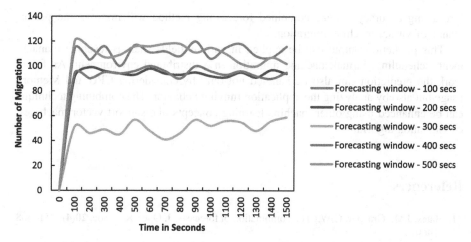

Fig. 7. Number Migrations for Simulation period of 1400 s by varying Forecasting Window W

generate Cloudlets by modifying the utilization model. The sudden spikes in workload has been simulated using random workload generator. The random workload generator introduce sudden spikes in the google workload in random interval of time. The simulation was carried out for 1400 s. The number of virtual machines is selected as 200 and stochastic utilization model is used. Space shared scheduling is used to schedule VM to the servers. The forecasting window W is varied to analyze the number of migrations in the cloud environment. The number of migrations for every 200 s is recorded by varying number of forecasting window and depicted in Fig. 7

It is observed that when the forecasting window is 100 s the server cluster has to reform more frequently. The random workload generator introduce random spikes that cause more number of migration. The condition was better in forecasting window 200 s. However the number of migration is greater than 100 in certain periods and this degrade the performance of the cloud service provider. The number of migration was considered to be optimal when forecasting window was fixed as 300 s.

It is also observed that the as forecasting window size increases above the optimal range the number of migrations increases. This condition occurs due to the random and dynamic workload spikes introduced. Due to random spike the server farms are not updated for certain period of time and cause severe migration flow in the cloud Datacenter. From the above simulation it is clear that the forecasting window should not be neither small nor too large. However dynamic forecasting windows can be used to improve the performance of the migration.

6   Conclusion

In this paper a novel methodology to predict the dynamic nature of workload in cloud datacenter has been proposed. The proposed methodology has been simulated and the results were obtained. The results were compared with the classical forecasting methods. The experimental results show that proposed methodology improves the forecasting accuracy. Hence combined forecasting method will improve the performance of virtual machine migration.

This prediction output can be applied in various fields such as resource management, scheduling, virtual machine migration in a distributed environment. As future work the prediction can also be handled not only by considering CPU and Memory usage but also by analyzing the application running behavior. The combining technique can be enhanced using other machine learning concepts like support vector machines, classification algorithm.

References

1. Bates, J.M., Granger, C.W.: The combination of forecasts. J. Oper. Res. Soc. **20**(4), 451–468 (1969)
2. Clark, C., et al.: Live migration of virtual machines. In: Proceedings of the 2nd Conference on Symposium on Networked Systems Design and Implementation, vol. 2. USENIX Association (2005)
3. Calheiros, R.N., et al.: CloudSim: a toolkit for modeling and simulation of cloud computing environments and evaluation of resource provisioning algorithms. Soft. Pract. Experience **41**(1), 23–50 (2011)
4. Ferreto, T.C., et al.: Server consolidation with migration control for virtualized data centers. Future Gener. Comput. Syst. **27**(8), 1027–1034 (2011)
5. Granger, C.W.J., Ramanathan, R.: Improved methods of combining forecasts. J. Forecast. **3**(2), 197–204 (1984)
6. Leelipushpam, G.J., Sharmila, J.: Live VM migration techniques in cloud environment—a survey. In: 2013 IEEE Conference on Information and Communication Technologies (ICT) (2013)
7. Hirose, Y., Yamashita, K., Hijiya, S.: Back-propagation algorithm which varies the number of hidden units. Neural Netw. **4**(1), 61–66 (1991)
8. Hines, M.R., Gopalan, K.: Post-copy based live virtual machine migration using adaptive pre-paging and dynamic self-ballooning. In: Proceedings of the 2009 ACM SIGPLAN/SIGOPS International Conference on Virtual Execution Environments. ACM (2009)
9. Nowotarski, J., et al.: Improving short term load forecast accuracy via combining sister forecasts. Energy **98**, 40–49 (2016)
10. Padullaparthi, V.R., et al.: Method and system for adaptive forecast of wind resources. U.S. Patent No. 9,269,056, 23 Feburary 2016
11. Strijbosch, L.W.G., Heuts, R.M.J., Van der Schoot, E.H.M.: A combined forecast—inventory control procedure for spare parts. J. Oper. Res. Soc. **51**(10), 1184–1192 (2000)
12. Van Ooyen, A., Nienhuis, B.: Improving the convergence of the back-propagation algorithm. Neural Netw. **5**(3), 465–471 (1992)

QOS Affluent Web Services Message Communication Using Secured Simple Object Access Protocol (SOAP) Technique

N. Anithadevi[1(✉)] and M. Sundarambal[2]

[1] Department of Computer Science and Engineering and Information Technology, Coimbatore Institute of Technology, Coimbatore 641014, India
anitha.cit@gmail.com
[2] Department of Electrical and Electronics Engineering,
Coimbatore Institute of Technology, Coimbatore 641014, India
sundaradurai@gmail.com

Abstract. In IT services SOA is one of the most elastic and modular approaches and it is a prerequisite for arising technologies like cloud, these cloud services are exposed as web services based on industry standards, which follows WSDL for service illustration. These services depend on SOAP to handle service request and response. Hence Web services security is one of the important factors which are used to assess cloud system security. While creating a new web service or with an existing web service communication, it is prudent to have secure data transmission with end users identity such as card numbers, user names, passwords etc. Security standards like WS-Security only addresses message integrity, confidentiality, user authentication, and authorization. The proposed system offers confidentiality and integrity protection from the creation of the message to its consumption. This system will look at a color palette scheme which records the RGB color values of the chosen color during registration and these values are used during sign on, subsequently it performs the access control mechanism. To strengthen web services towards message level by encrypting SOAP messages with AES and shared key is derived using new cryptosystem called Rbits (Random bits) cipher as a service and digital signature handler facilitates secure key exchange which is completed ahead with SOAP message generation. The essential aspect of this proposed system is from core key form multiple random keys which safeguards the messages with highest possible immunity to crack when the applications or services communicating with web services.

Keywords: Color palette scheme · Simple Object Access Protocol (SOAP) · AES (Advanced Encryption Standard) · Random bits (Rbits) · Digital signature

1 Introduction

Broader application accessibility is achieved when widening the Web services concept. But it is also catalysis for collaboration between several distributed applications through the composition concept. The security breaches and threats which are exposed by that

© Springer Nature Singapore Pte Ltd. 2016
S. Subramanian et al. (Eds.): CSI 2016, CCIS 679, pp. 191–207, 2016.
DOI: 10.1007/978-981-10-3274-5_16

web services should be effectively handled for its secured communication. Based on the users need, Web services are created and with their preference and requirement, they are composed with other web services. Web services almost exclusively use SOAP (Simple Object Access Protocol) for all messaging. The first important protection in this type of communication is during wired communication where, the SOAP messages that are sending across the services have to be delivered confidentially and without tampering. The second focus is that it has to ensure that the content of the communications can only be read by the right server, and also it should ensures that it can only be read by the right process on the server and each one has to be signed with proper message level security. After that it has to focus System message logs to reconstruct the chain of events with adequate information to track those back to the authenticated callers.

This paper proposes an infrastructure that supports the dynamic composition [1] and secure communication among web services. The first and foremost requirement is User authentication when a Web service is being consumed by other Web developers or by the general public, which leads to the next level of security called secure message communication, that is messages containing sensitive data must be encrypted with strong data encryption, not just transport encryption. Using strong encryption technique, both the sender and the receiver have to encrypt their sensitive data. A significant quality of this kind of message focal is hasty encryption with huge amount of data with restricted resource and time by the sender. In the receiving end, the receiver faces many forgery attempts with large volumes of data. In order to overcome these complications, the sender and receiver encrypt messages using strong encryption function that should handle large volumes of data with adequate encryption time which is also suitable for wider range of applications. Web services security can be classified in to two major categories. First one is to how to protect the confidentiality and integrity of a message and the second one is to employ the measures to guarantee that authentic messages are received only from authorized parties which are considered below.

2 Related Work

In this section, we will review some existing schemes which provide secure web services composition and communication. This helps to turn core security features such as Authentication, Authorization, Auditing, Confidentiality, Integrity, and Availability into action.

2.1 Color Scheme Authentication

In Web services, user authentication is one of the common techniques to restrict unauthorized access and usages of services. Textual password is the prevailing technique used for user authentication, but eves dropping, dictionary attack, social engineering and shoulder surfing vulnerabilities of this kind of approach makes it unreliable. Graphical passwords and biometrics are other alternative techniques which also have their own disadvantages like slow identification process and expensive nature. In registration phase, users are required to choose their pictures, symbols or icons from a collection in Recognition-based schemes and also with a recognition phase of their choice for successful identification.

An authentication mechanism must provide adequate security in order to meet its goal. Guessing attacks (brute-force and dictionary attacks) and capture attacks (shoulder surfing, spyware and social engineering attacks) are the major broad attacks sections of graphical password system. RGB is a general colors used in web page design. Netscape Color Cube [1] defines 216 RGB colors as the limited color depth of most video hardware.

2.2 Web Services Security

In distributed computing environment, security is one of the Influential factor which leads to secure communication. Using web services, businesses percolate many transactions among their trading partners. In the frame of reference Internet communication is not a great defender against intruders. To overcome this issue equips policies and mechanisms to emphasize Web services security. Numerous Security standards such as Assertion Markup language (SAML) [2], WS-Security [3] and WS-XACML [4] were proposed for secure communication. WS-Security is nothing but the SOAP messages are appropriately encrypted and digitally signed using XML. In today's e-business scenario, organizations are investing a huge amount of their resources in Web Services and apparently they complete their transactions using plain-text, XML formats like SOAP and WSDL which leads to effortless hacking. To safeguard XML documents XML Signature and XML Encryption plays major role and also appropriate maintenance of document structure and smooth implementation. This can be estimated using the parameters authentication, authorization, integration, confidentiality, and non-repudiation.

Elemental web architecture and HTTP transport protocol build common Web Services which leads to identical threats and Vulnerabilities in distributed environment. To safeguard web services from intruders, Web Service Security (WS-Security) is one of the impressionable and feature-rich extension to SOAP based web services and to regulate this, numerous security standards are created. In the time of service provider and consumer message communication, innumerable SOAP messages are exchanged among them which escalate the security overhead. To overcome the security overhead, users avail XML encryption, decryption and signatures with increased CPU resources and WSS also give security extensions which are established using PKI, X.509, Kerberos, or other algorithms which also yield increased overhead [15]. Users can request and use web services far and wide with existing security standards and there can be a security hole if it offers for a specific domain or enterprise which necessitates a new secure token to verify, control or monitor the service consumer location [13]. With this recommended token sender location can be validated which slashes demand on system resources by holdup bogus users.

2.3   SOAP-Based Web Service Security

In web services architecture, SOAP is the most prevailing core component, in fact it is a specification for transferring structured messages among computers through common communication protocols. SOAP message formats build based on XML. Communication protocols like Hyper Text Transfer Protocol (HTTP), Simple Mail Transfer Protocol SMTP give foundation to implement interoperable machine-to-machine interaction over the Internet and also with this the interaction can also pass through the firewall. In SOAP, message level security and transport level security are the two important layers for secure communication. To protect the data, such as IPSec, SSL, etc. the transport level security utilizes layer 3 or layer 4 protocol and Layer 7 is responsible for message level security and that has to be protected efficiently for end to end communication.

2.4   SOAP Message Security

In order to attain secure message exchanges among web services, message integrity and message confidentiality are the two major objectives. To achieve these objectives "WSS: SOAP Message Security" is a kind of pattern which leads to construct secure Web Services With the help of XML Signature [11] XML Encryption [12]. The SOAP message consists of SOAP header (optional element) and SOAP body (mandatory element) which encompasses value-added services, payload or information, respectively to transfer and illustrate the data XML-based message elements or tags are used also these tags should be acknowledged and accepted among the participants. To safeguard SOAP messages during web services communications we can use Web Service Security (WSS) Specification which uses <wsse:Security> tag to append security related information. With this WSS foundation recipient can determine message security information and also recover the details of the processing rules. Here, message encryption and signature are based on the XML-based security tokens which are used to provide an authentication and authorization during communication. Helander and Xiong came up with secure web service framework based on peer-to-peer model [5, 6] This illustrates two-way authentication and peer-to-peer key exchange which is based on RSA and AES [7, 14].

Using RSA algorithm two-way authentication and peer-to-peer key exchange were achieved and with the help of AES algorithm transmission and communications were encrypted. Authentication and key establishment can be realized using either the symmetric key method or the public key method [7]. End- to- end security [8] solution for m-banking in wireless networks utilizes Advanced Encryption Standard (AES) to ensure data confidentiality, authentication, and data integrity also maintained effectively.

Symmetric and Asymmetric key encryption algorithms are major two classifications encryption algorithms which uses same key used for encryption and decryption [10] and different key used for encryption and decryption, respectively. Encryption algorithms like AES, DES, 3DES, BlowFish and IDEA [10–12] fall on to symmetric key encryption algorithm category and the significant favor for using these symmetric algorithms are its rapid speed and security [16–18]. Based on performance and security

AES is relatively one of the efficient encryption techniques than others. To achieve 128 bit key symmetric encryption algorithms level of security, Asymmetric encryption algorithms have to use 3,000-bit key and it is slow and not feasible to handle huge amount of data. Generally they are used concurrently where to encrypt a randomly generated encryption keypublic-key algorithm is used and with the help of that random key we can encrypt the actual message using a symmetric algorithm which is labeled as hybrid encryption [19].

3 Method and Analysis

In this proposed model, a color Scheme Authentication pattern is used to validate the user when they get in to the services. This pattern is a kind of session passwords that should use only once that is for each and every login it acquire distinct password that is exclusively for a single time. In this method, user can give text or numbers to choose colors. This method safe guards the service from shoulder attack, dictionary attack, eves dropping etc. In 24-bit displays, the use of the full 16.7 million colors of the HTML RGB color code no longer poses problems that is color palette consists of the 216 (6^3) combinations of red, green, and blue have six values (in hexadecimal): #00, #33, #66, #99, #CC or #FF based on its intensity it may represent 0%, 20%, 40%, 60%, 80%, 100%, correspondingly. In this way 216 colors are splitting up into a cube of dimension 6. CRT/LCD's standard 2.5 gamma perceived intensity is only 0%, 2%, 10%, 28%, 57%, 100%. This type of Web color Palette is used in this model to change the color of images to authenticate users from unauthorized access.

To acquire data integrity in web services, disparate encryption algorithms like (DES, 3DES, AES etc.) are employed to safe guard SOAP messages during web services communication. To achieve remarkably secure web service communication, the security level of each one of these ciphers is evaluated to conclude highly secured web service communication. With respect to this we propose AES encryption algorithm to safe guard SOAP messages and with Rbits security method shared key is generated, again this key exchange is protected using SHA1 digital signature to attain utmost protection. With the help of AES algorithm Service requester and the service provider should encrypt and decrypt the SOAP messages, respectively in order to obtain its original message. AES or Advanced Encryption Standard is a symmetric key encryption technique of 128 bit block size, and its key size ranges between 128, 192 and 256 bit. Based on its key sizes, the number of round operations varies (i.e. AES128 10 rounds, AES192 12 rounds AES256 14 rounds). AES algorithm Encryption phase is divided in to 2 phases such as Key Expansion (Expands the key into the required number of keys) and Message Encryption (data block goes through n rounds depending on the key size). Each round consists of a SubByte substitution, ShiftRow, MixColumn and Add round key.

With the foundation of Michael O. Rabin's oblivious transfer mapping, Lenore Blum, Manuel Blum and Michael Shub.

Proposed BlumBlumShub (B.B.S.) pseudorandom number generator in 1986 which has the following format.

$$x_{n+1} = x_n^2 \bmod M \tag{1}$$

$$x_i = \left(x_0^{2i \bmod \lambda (M)}\right) \bmod M \tag{2}$$

Here,

M = pq where p and q are two large primes.
x_{n+1} = Each step Output.
x_0 = co-prime to M (i.e. p and q are not factors of x0) and not 1 or 0 it must be integer.

In this, some output is got from x_{n+1} or the bit parity of x_{n+1} or one or more of the least significant bits of x_{n+1}. The two large primes, p and q, must be congruent to 3 (mod 4) (this guarantees that each quadratic residue has one square root which is also a quadratic residue) and gcd (φ(p–1), φ(q–1)) should be small (used to create huge cycle length) and using Euler's theorem Blum BlumShub generator can have a option to calculate any x which is its remarkable feature. Penchalaiah and Ramesh Reddy proposed Rbits (Random bits) a symmetric key block stream cipher in 2013 which generates random bits at sender and the receiver side which is used to encrypt and decrypt and these random bits are created based on Blum BlumShub (BBS) algorithm.

4 Proposed Technique/System

Figure 1 illustrates main architecture flow of the proposed technique. Securing the SOAP messages during web service communication is the main target of this proposed model and this web service model consists of set of components which step up the secure SOAP message transmission than the conventional web service models. This Rbits keybased SOAP message security model is discussed in this paper.

4.1 Key Value Generation

The public key and privatekey pair values which are required to the requestor and the provider during service communication were created in this module and using requester and provider's credentials they are used to obtain their public key. These keys are digitally signed and verified and it transferred using Rbits, which is created via Blum BlumShub (BBS) algorithm. Service Requester (SR) and the Provider (SP) should use BBS algorithm. Common shared Core-Key (SCK), parameter 'P' are the inputs to BBS to create random bits and the most important part of this algorithm is randomly changing multiple keys (RKi) (Bellare et al. [9]). This is used for encryption using a single key SCK and P which are common to both requester and Provider. The significant feature of BBS generator is it is unpredictable to left and right (i.e. the cryptanalysts couldn't calculate the next bit or the previous bit from the BBS generated bit sequence) and direct manipulation of x values.

$$x_i = \left(x_0^{(2^i (mod((P-1)*(Q-1))))} \right) (modN) \tag{3}$$

During service communication, it require multiple keys in the sense that can be calculated and recovered from previous values and the initial x and N (Junod [21], Blum *et al.* [20]) which preserves memory (i.e. No need to store all keys).

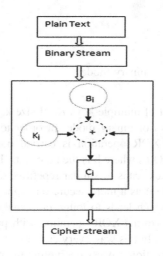

Fig. 1. Rbits encryption process

Where

Bi - ith Plain Block Stream
+ - XOR operation
Ki - ith Key
Ci - Cipher Block Stream

4.2   Deriving Common Shared Core-Key (SCK) and Parameter (P)

The common Shared Core-Key (SCK) is retrieved from BBS.
  Preliminary steps as follows

1. Select two large prime numbers, x and y. $x \equiv y \equiv r$ (mod m).
2. Manipulate $n = x * y$.
3. Choose a random number 's', relatively prime to 'n'
4. Calculate $Z_0 = s2\ mod\ n$. $SCK = Z_0$

  Here 'Z_0' is the common shared core key 'CK' and 'n' is parameter 'P'.

4.3   Multiple Key Generation with CBC Mode

With the help of BBS algorithm multiple keys are created the same as follows.

$loopi = 1$ to msg_{len}
$sck_i = (sck_{i-1})2 \ mod \ n$
$b_i = sck_i \ mod \ 2$
$kj = kj\| \ bi$
if $(i\% \ k_{size}) = 0$ then
$j = j + 1$
end loop

msg_{len} = Message length in binary mode.
k_{size} = Key length.

With an adequate amount of multiple keys Ki of size k_{size} are created to encrypt a message with msg_{len} instream or binary mode. In order to build this algorithm as a trouble-free and faster one XOR operation is carried out and its data processing complexity is represented as O(b) where b is the no of bits. In this algorithm, repetitive pattern of same plaintext block leads to similar repetitive ciphertext blocks. To overcome this difficulty and make it as a more secure one append Cipher Block Chaining (CBC) with this algorithm which leads to tough cryptanalysis. In this CBC mode of operation plain text block performs XOR operation with preceding cipher text block prior to its encryption process. In this way every cipher text block is depend with other plain text blocks. In the decryption process decrypting the present ciphertext and then adding the preceding ciphertext block leads to the final resultant message. This type of operation is suitable for services which have a need of both symmetric encryption and data origin authentication. During cryptanalysis statistical properties of the on hand ciphertext is utilized. Here this model protects the statistical characteristics of the plaintext with the additional CBC mode of operation which defends from any type of cryptanalysis (Bellare *et al.* [9]). In 32-bit platform, bit which is less than 2 GB of addressable space are in process and some of the remaining are fragmented into smaller pieces. Unless the message size of its default setting is increased, the maximum size of message will be 64 KB. So, the proposed work is tried with up to 64 KB with Base 64 encoding. If the Base 64 results in larger messages, the message is compressed with the help of WSE filter tools, such as GZIP. The SOAP Messages size is limited to an appropriate size limit (here 64 KB) because Larger size limit (or no limit at all) increases the chances of a successful DoS attack

4.4   Rbits Encryption and Decryption Handlers

Both the sender and the receiver must fulfill the pre-requisites before the encryption and decryption process which is illustrated as follows.

(a) A strong Security Binding (SB) must be formed among the sender and the receiver which leads to proper core-key exchanging, parameter agreement(parameter here means block size).

(b) Same Block size should be used for both sender and receiver.
(c) Core-key, Parameter and block size should be kept secret and infeasible to predict.

Encryption Handler. During Encryption process the sender or requester has to perform the following operations.

(a) Given message has to be converted in to stream of bits.
(b) The binary stream is partitioned into definite block sizes called block stream.
(c) To create random block of bits at sender end, acquire a block stream and XOR with key Ki.
(d) Utilize CBC encryption operation.
(e) Repeat Step c and d till last block stream of message produces cipher stream.

CBC encryption process is illustrated as follows

$$C_i = E_k (p_i \oplus C_i) \tag{4}$$

$C_0 = IV$
IV = Initialization Vector.

Decryption Handler. During Decryption process the, Receiver or provider has to perform the following operations.

(a) The cipher stream is converted in to specific size of block of bits as sender.
(b) Acquire a block stream and XOR with key Ki with instantly generated random block of bits at receiver or provider end.
(c) Utilize CBC decryption operation.
(d) Repeat Steps b and c till last block stream of message.
(e) Convert stream of bits in to plaintext.

$$P_i = D_k(C_i \oplus C_{i-1}) \tag{5}$$

$C_0 = IV$

Signature Creation. When a web service is evoked, with the above security handlers the service is acquired safely. In order to safeguard it from intruders the service requester and provider wrap the above public value which is generated using Rbits with digital signature as follows.

The requester carries out the public value creation as described above (Rbits) and generates the signature for a requester and sends the Rbits generated public value using Digital Signature produced to the provider.

$$SCK + D_{Sig}(SCK, WS_{Req}) \tag{6}$$

Then the provider detaches the digital Signature $D_{Sig}(SCK,WS_{Req})$ and the public key value SCK and verifies the digital signature if it matches then it proceeds its communication otherwise it declines its further process. In the same way the providers also performs the digital signature formation and verification operation and safeguard its service communication.

$$SCK + D_{Sig}(SCK, WS_{Pro}) \tag{7}$$

In web services both Requester and the Provider have to verify their identity using Certificate Authorities. In the proposed work, the provider and requester credentials are verified using trusted authority such as a digital certificate (Secure Hashing Algorithm SHA1). Once the signature is verified, the requester or the provider can ensure the message is not tampered. They can also validate the certificate to make sure that it is not expired or revoked and ensure that no one has actually tampered their private key using the Formulas 6 and 7 which leads to trusted message communication.

4.5   SOAP Message Encryption Algorithm

When requester or the provider acquire the web service for its services throughout its communication the message from Web service SOAP request and the message obtained from the client is safeguarded using AES encryption algorithm as proceeds.

SOAP Request Encryption and Decryption Handler. During web service communication when the service is requested then the SOAP request encryption handler obtain the message from SOAP request and the message received from the client is encrypted using AES algorithm by utilizing the shared core key SCK. Then the encrypted message ($AES_{encrypt}(SCK,message)$) is safely transmitted to the provider. To decrypt that message we have to depend on the SOAP decryption handler to acquire the encrypted message from the SOAP request and decrypts it using AES algorithm with the generated secret key.

SOAP Response Encryption and Decryption Handler. Throughout web service communication SOAP Response Encryption Handler handles SOAP response messages and encrypts the same using AES with its secret key. This Encrypted SOAP response messages will be decrypted via SOAP Response decryption Handler. The encrypted message which is received from response handlers ($AES_{encrypt}(SCK,message)$) is decrypted using ($AES_{decrypt}(SCK,message)$). Here, for each block of stream distinct key is used and its total Key length is as long as the length of the SOAP message and it achieves unpredictable randomness of Key for each Block with its added CBC mode of operation which makes complex cryptanalysis.

AES Encryption. At the time of web service communication, AES encryption and decryption is carried out with the help of above handlers in both web service requester and provider region. AES algorithm works with 4×4 column-major order matrix of bytes, for the plain text with 128 bytes and it is composed with the following four round of operations.

1. SubBytes
2. ShiftRows
3. MixColumns
4. AddRoundKey

Initial round starts with add round key and the Mixcolums operation is not there in the final round. The number of rounds/cycles are replicated based on its key size, 10 cycles of repetition for 128-bit keys, 12 cycles of repetition for 192-bit keys and 14 cycles of repetition for 256-bit keys. Each round of operation is performed with the above mentioned steps with the help of encryption key and set of reverse rounds are carried out to convert ciphertext back into the original plaintext using the same encryption key [11].

This proposed secure web service communication model performs its service communication safely as illustrated in the following Fig. 2.

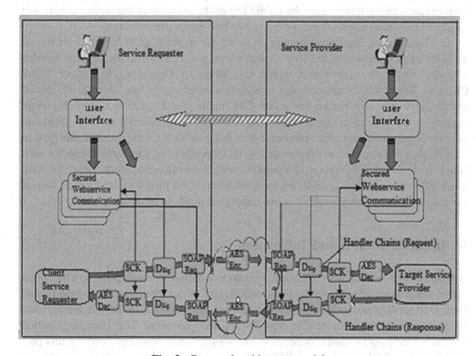

Fig. 2. Proposed architecture model

When a service requester requests a web service through the user interface then it sends the request to the proposed secure communication channel, where the client service requester and provider communicate safely. Initially, a requester selects a random private key and create a public key with the help of Rbits encryption and decryption handlers for each block and send it to signature handlers, there it creates digital signature using requesters private key and wrap this with signature conjointly with its public key and forward the SOAP request to the provider. With the help of SOAP message Encryption, handlers encrypt the SOAP message using AES encryption Algorithm.

Service provider acquires the request received from requester and separates the message and its signature. Initially, it performs signature verification using requesters public key and it verifies the signature, once it matches it proceeds otherwise it denies the request, with this we achieve data integrity. By SOAP message Decryption handlers decrypt the SOAP message using AES decryption Algorithm. In the meantime Rbits Encryption and decryption handlers choose random number and creates public key (for each block) which is send to the provider. Then, the signature handlers create digital signature using the private key of the provider and enfold the message with this signature and send it as response to the requester. To enhance the security here also the SOAP encryption handlers are called to encrypt the SOAP response message. Once the requester receives the request which is transmitted by the provider then the same process repeats as did in provider side like message and the signature separation and verification with signature using the public key of the provider to maintain integrity and finally SOAP decryption to receive Secure SOAP Response messages.

In block cipher, the transformation is invertible. It maps an N bit block of bits to an N bit block of bits with the Key control where n = 128 in the case of AES and this mapping blocks are performed within the Mode of Operation. In Cipher Block Chaining Mode (CBC) the message is always multiple of N bits, so padding can be added to the message before the actual CBC-mode transformation is done for more security and it is hard to cryptanalysis. AES with Rbits is suitable for both short and long message communication, moreover it is faster in both Encryption and Decryption Process (but it is complex to cryptanalysis). Its Unpredictable Key randomness for each Block with added CBC mode of operation satisfies avalanche effect. Here, for each block of stream new key is used. All keys are distinct and total Key length is as long as the length of the plaintext which is not supported much in other algorithms.

5   Implementation and Inferences

A hospital web service is created and a separate database is allotted for it. The doctor names are obtained by giving the place and disease as the input. Similarly blood bank, bank web service and a travels web services are also created. Bank web service performs functions like withdrawals deposits and bill payments. The User Registration system which possess credit/Debit card for bill payments is safeguarded by this proposed model. With the help of Rbits key exchange, the key is shared among the requester and the provider safely with digital signature wrap and using AES algorithm the SOAP message during request and response is encrypted at each end only when both requester and provider are trusted (with proper key exchange) as represented in the Fig. 5.

The Utility ratio of the system is interpreted as the ratio of the SOAP messages with the digitally signed key exchange with the adequate procurement of SOAP message communication with authorized users measured in terms of %.

$$UR = \frac{(No.\ of\ SOAP\ messages - Retrival\ Rate\ of\ SOAP\ messages)}{(No.\ of\ SOAP\ messages)} \quad (8)$$

Table 1. Utility ratio

Document size (MB)	AES-Rbits	AES-Diffie–Hellman
5	55	50
10	59	57
15	63	60
20	61	63
25	65	66
30	67	68
35	72	70
40	74	71

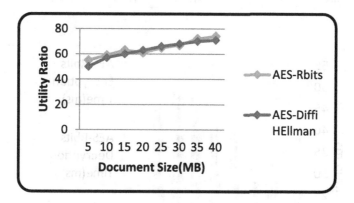

Fig. 3. System utility ratio

This Secured Key exchange with AES encryption algorithm yield higher utility ratio when compared to the formal Diffie Hellman key exchange. Table 1 illustrates the proposed algorithm offers tremendous utility ratio when compare to formal Diffie Hellman. The Rbits key exchange with digital signature performed using hash values and random multiple keys for each block commence increased utility ratio of a system as shown in Fig. 3. To ascertain the performance of the decryption time, for numerous runs the time taken by the system is explored for AES with Diffie Hellman and AES with Rbits key exchange and illustrated as follows. Decryption time Using this proposed Secured AES algorithm to respond the provider or requester is measured in terms of milliseconds (Fig. 4).

$$DT = Time(Enhanced\ AES(Rbits(D_{Sig} + CPU\ time)))\qquad(9)$$

The above graphical representation illustrates that the decryption time is minimal when compared to Diffiehellman key exchange because of its variable length of block sizeand random multiple key structure. This analysis simulation is performed on

Table 2. Speed analysis (encryption & decryption)

Doc Size (MB)	AES-Rbits		AES-Diffie Hellman	
	Encryption Time (ms)	Decryption Time (ms)	Encryption Time (ms)	Decryption Time (ms)
5	16	15	20	18
10	32	16	31	26
15	32	30	34	29
20	32	41	37	44
25	34	43	37	45
30	39	46	42	48
35	43	48	45	50
40	42	47	47	51

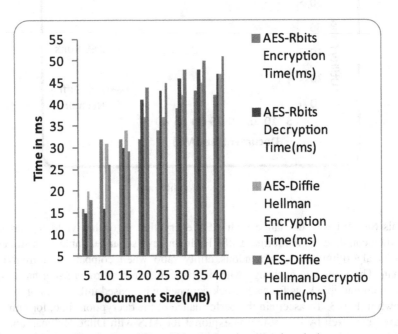

Fig. 4. Encryption & decryption time speed analysis

windows 7 OS, 3.0 GHz Dual core intel processor, using Java 1.6 and IDE Netbeans. This interpretation builds upon the underlying operating system, and Hardware resources available at that time and this algorithm is applied on 5 MB to 40 MB of data and the experimental results are illustrated in Figs. 5 and 6 while communicating with Hospital web service Environment (Table 2).

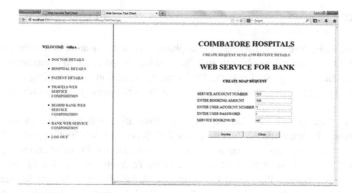

Fig. 5. Web service invocation

Fig. 6. Cipher text processing and signature verification

6 Conclusion and Future Work

An AES based SOAP message security with Random multiple key values have been constructed, to attain security on SOAP message communication and to enhance the security, protect the key exchanges with digital signatures and color palette scheme during sign on. Subsequently, it performs the access control mechanism. In addition to this, during web service communication either in service request or response, the SOAP messages in service communication are separated in to different blocks and for each block unique random multiple key is generated and prevent it from transmission overheads. This multiple keys to each block of message ensure favorably secured ciphertext and the added CBC mode process conceal the statistical characteristics of the messages to some extend which leads to complex cryptanalysis.

References

1. Bonnefoi, P.-F., Xydas, I., Krikelas, I.: Graphical user authentication in mobile device using the web RGB color palette. In: 6th BCI 2013-Balkan Conference in Informatics (2013)
2. Chen, S., Zic, J., Tang, K., Levy, D.: Performance evaluation and modeling of web service security. In: IEEE International Conference on Web Services (ICWS 2007), pp. 431–438 (2007)
3. Chen, S., Yan, B., Zic, J., Liu, R., Ng, A.: Evaluation and modeling of web services performance. In: International Conference on Web Service (ICWS 2006), pp. 437–444 (2006)
4. Xiong, K.: Web services performance modeling and analysis. In: International Symposium on High Capacity Optical Networks and Enabling Technologies, pp. 1–6 (2006)
5. Mondejar, R., Garcia, P., Pairot, C., Skarmeta, A.: Enabling wide-area service oriented architecture through the p2pWeb model. In: 15th IEEE International Workshops on Enabling Technologies Infrastructure for Collaborative Enterprises, pp. 89–94, June 2006
6. Helander, J., Xiong, Y.: Secure web services for low-cost devices. In: IEEE International Symposium on Object-Oriented Real-Time Distributed Computing, 18–20 May 2005, pp. 130–139 (2005)
7. Stallings, W.: Cryptography and Network Security: Principles and Practices, 4th edn. Prearson Education, Upper Saddle River (2006)
8. Hassinen, M., Laitinen, P.: End-to-end encryption for SMS messages in the health care domain. Stud. Health Technol. Inf. **116**, 316–321 (2005)
9. Bellare, M., Cash, D., Keelveedhi, S.: Ciphers that securely encipher their own keys. In: 18th ACM Conference on Computer and Communications Security, pp. 423–432, October 2011
10. Verma, O.P., Agarwal, R., Dafouti, D., Tyagi, S.: Performance analysis of data encryption algorithms. In: 3rd International Conference on Electronics Computer Technology (ICECT) (2011)
11. Alanazi, H.O., Zaidan, B.B., Zaidan, A.A., Jalab, H.A., Shabbir, M., Al-Nabhani, Y.: New comparative study between DES, 3DES and AES within nine factors. J. Comput. **2**(3), 152–157 (2010)
12. Seth, S.M., Mishra, R.: Comparative analysis of encryption algorithms for data communication. IJCST **2**(2), 292–294 (2011)
13. Cheong, C.P., Chris, C., Young, R.: A new secure token for enhancing web service security. In: 2011 IEEE International Conference on Computer Science and Automation Engineering (CSAE), vol. 1. IEEE (2011)
14. Beimel, A.: Secure schemes for secret sharing and key distribution. Ph. D. dissertation, Israel Institute of Technology, Technion City, Haifa, Israel (1996)
15. Chung, K.-M., Kalai, Y., Vadhan, S.: Improved delegation of computation using fully homomorphic encryption. In: Rabin, T. (ed.) CRYPTO 2010. LNCS, vol. 6223, pp. 483–501. Springer, Heidelberg (2010). doi:10.1007/978-3-642-14623-7_26
16. Damgård, I., Thorbek, R.: Linear integer secret sharing and distributed exponentiation. In: Yung, M., Dodis, Y., Kiayias, A., Malkin, T. (eds.) PKC 2006. LNCS, vol. 3958, pp. 75–90. Springer, Heidelberg (2006). doi:10.1007/11745853_6
17. Qin, B., Wu, Q.H., Zhang, L., Farras, O., Doming-Ferrer, J.: Provably secure threshold public key encryption with adaptive security and short ciphertexts. Inform. Sci. **200**, 67–80 (2012)
18. Zhang, Z., Zhu, L., Liao, L., Wang, M.: Computationally sound symbolic security reduction analysis of the group key exchange protocols using bi-linear pairings. Inform. Sci. **209**, 93–112 (2012)

19. Waleed, G.M., Ahmad, R.B.: Security protection using simple object access protocol (SOAP) messages techniques. In: International Conference on Electronic Design, ICED 2008. IEEE (2008)
20. Blum, L., Blum, M., Shub, M.: A simple unpredictable pseudo-random number generator. SIAM J. Comput. 364–383 (1986)
21. Junod, P.: Cryptographic secure pseudo-random bits generation: the Blum-Blum-Shub generator (1999)

A Reactive Protocol for Data Communication in MANET

M. Anandhi[1($\boxtimes$)], T.N. Ravi[2], and A. Bhuvaneswari[1]

[1] Cauvery College for Women, Trichy, India
anandhia@yahoo.com, prkrizbhu@yahoo.co.in
[2] Periyar E.V.R College (Autonomous), Trichy, India

Abstract. A MANET (Mobile Ad-Hoc Network) is composed of mobile, autonomous, wireless nodes that could be connected at network edges to that of fixed wired internet. The ad-hoc network does not need infrastructures that are required for other wireless networks. The term infrastructure includes need of base stations, routers etc. The network execution ought to be supported up by protecting collaboration among various nodes and it is finished by essential mechanisms. In further, most proficient routing protocols are utilized to route the packets to the destination from the source by the method of routing process. In MANET, there are various kinds of routing protocols and every one of them is associated with systematic circumstances. This paper focuses on designing a new routing protocol which considering minimum number of nodes for routing and select optimal route. This protocol also concentrates clustering as well as security considerations in order to provide tenable and proficiency routing.

Keywords: MANET · Routing · DTV · On-demand · Prophecy · Cluster

1 Introduction

The chief characteristics and challenges of the MANETs are *Cooperation*: MANETs rely on the cooperation of the nodes for routing and packet transmission. If the source and destination nodes are not in the range of each other then the communication between them takes place with the cooperation of other nodes. All the nodes between them form an optimum chain of mutually connected nodes. In this each node is to act as a host as well as a router. Simultaneously, this is known as multi hop communication. *Dynamism of Topology:* The MANET nodes are random and unpredictable and so is the topology. The nodes may leave or join the network at any point of time also the topology is vulnerable to link failure, all these affect the status of trust among nodes and the complexity of routing. *Lack of fixed infrastructure:* The absence of a fixed or central infrastructure is a key feature of MANET. There is no centralized authority to control the network characteristics. Due to the absence of authority, traditional techniques of network management and security are scarcely applicable to MANET. *Resource constraints:* MANET is a set of mobile devices which are of low or limited power capacity, computational capacity, memory, bandwidth etc. by default. So in order to achieve a secure and reliable communication between nodes, these resource constraints make the task more enduring.

S. Subramanian et al. (Eds.): CSI 2016, CCIS 679, pp. 208–222, 2016.
DOI: 10.1007/978-981-10-3274-5_17

In MANET, there are various kinds of routing protocols and every one of them is associated with systematic circumstances. The routing protocols for ad hoc networks can be broadly classified into four categories based on Routing information update mechanism, and based on the usage of temporal information for routing. Routing topology and Utilization of specific resources.

Proactive, Reactive and Hybrid routing protocols are three major categories based on the routing information update mechanism. The protocols based on the use of temporal information for routing can be further classified into two types: Routing protocols using past temporal information and future temporal information. There are two other major protocol types which are based on routing topology: Flat topology routing protocol and Hierarchical topology routing protocols. Based on the Utilization of specific resources the routing protocols can be classified as Power-aware routing and Geographical information assisted routing. Proactive Routing Protocol: Proactive routing protocols are also known as table driven routing protocols. In this routing mechanism every node keeps up routing table which restrains information in relation to the network topology yet without entailing it [1, 2]. This component albeit valuable for datagram movement, acquires the common signaling traffic and power utilization [3]. Reactive Routing Protocol: The routing protocol in which the route is initiated at any point where a path is required by a node to communicate with a destination. This kind of protocols do not maintain the network topology information. In addition, the source node perceives its route store for the available route from source to destination and if the route is not reachable then it institutes route discovery method [4].

2 Related Works

Destination Sequenced Distance Vector (DSDV) routing protocol [5] is one of the table-driven protocol where each node maintains a table that contains the shortest distance and the first node on the shortest path to every other node in the network. It incorporates table updates with increasing sequence number tags to prevent loops, to counter the count-to-infinity problem, and for faster convergence. The tables are exchanged between at regular intervals to keep an up-to-date view of the network topology. The tables are also forwarded if a node observes a significant change in local topology. Much less delay is involved in the route setup process. This protocol suffers from excessive control overhead and is not scalable. Another drawback of DSDV is that in order to obtain information about particular destination node, a node has to wait for a table update messages initiated by the same destination node.

The Dynamic Source Routing (DSR) [6, 7] is an on-demand protocol designed to eliminate the periodic table-update messages required in the table-driven approach. The routing construction phase is to establish a route by flooding *Route Request* packets in the network. The destination node on receiving a *Route Request* packet, responds by sending a *Route Reply* packet back to source. The intermediate nodes also utilize the route cache information efficiently to reduce the control overhead. This protocol lacks in route maintenance mechanism which does not locally repair a broken link. Route setup delay and routing overhead are some drawbacks of this protocol. The performance of this protocol degrades rapidly with increasing mobility.

Ad Hoc On-Demand Distance Vector (AODV) [8] is an on-demand approach which employs destination sequence number to identify the most recent path. The source node and the intermediate nodes store the next-hop information corresponding to each flow for data packet transmission. This protocol use destination sequence number (DestSeqNum) to determine an up-to-date path to the destination. A node updates its path information only if the DestSeqNun of the current packet received is greater than the last number stored at the node. The advantages of this protocol are it finds the latest route and connection setup delay is less. One of the disadvantages of this protocol is that intermediate nodes can lead to inconsistent routes. Multiple *Route Reply* Packets in response to a single *Route Request* can lead to heavy control overhead. Another drawback is that the periodic beaconing leads to unnecessary bandwidth consumption.

Temporally Ordered Routing Algorithm [9, 10] which uses a *link reversal algorithm* and provides loop-free multipath routes to a destination node. This process establishes a destination-oriented directed acyclic graph using *Query/Update* mechanism. By limiting the control packets for route reconfigurations to a small region, TORA incurs less control overhead. But local reconfiguration of paths results in non-optimal routes.

Location-Aided routing protocol (LAR) [11] utilizes the location information for improving the efficiency of routing. LAR designates two geographical regions *Expected Zone* is the region in which the destination node is expected to be present and *Request Zone* is the region within which the path-finding control packets are permitted to be propagated. The path-finding control packets are discarded by the nodes outside of the zone. LAR reduces control overhead by limiting the search area for finding a path and increased utilization of bandwidth. LAR cannot be used in situation where there is no access to location information.

3   Work Descriptions

The proposed method concentrates on designing a new protocol named as Dynamic Triangular Vision and Optimized Slant Selection Protocol (DTVOSSP) which uses the location information to improve performance of routing protocols. In this proposed

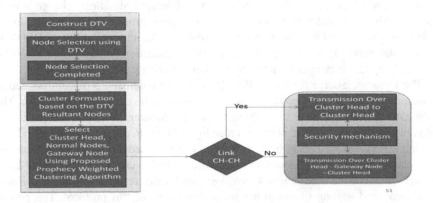

Fig. 1. The structure of the proposed method

protocol there are three modules which are names as Dynamic Triangular Vision (DTV) algorithm a new approach in route discovery, Prophecy weighted clustering algorithm which is based on DTV and Security algorithm. The overall structure is given Fig. 1.

3.1   DTV Algorithm

A Dynamic Triangular Vision algorithm is an on-demand routing scheme based on location information of nodes which are updated based on location management. This method reduces control overhead. When a node starts to communicate with a node, a triangular vision request zone is created between source and destination to reduce the route discovery region. The DTV mechanism shows the best way to select a route to the destination in on-demand basis. In this algorithm the source node does not send route request packet to all its neighboring nodes. The source node selects only few nodes within a triangular region which is focused towards the destination. This method uses flooding, but here flooding is restricted to a small geographical region. Within request zone only limited nodes can be elected based on some weights. Each node maintains a DTV table which contains the information of all its adjoining nodes likes node ID, location and Grade Point (GP).

$$GP = W_1X_1 + W_2X_2 + W_3X_3. \ldots \tag{1}$$

Where W_i represents weights of a metric i and $0 < W_i < 1$. X_i represents the value of metrics i which represents the node quality. The DTV table will be updated regularly and it is maintained in ascending order of GP values. The nodes which satisfying the constraints (within request zone) are considered as feasible nodes and top three nodes with highest GP values are treated as optimal nodes. The various metrics that are used in ranking the nodes are given below:

X_1 - *Packet Forwarding Ratio.* The cooperative node can be identified by their previous Packet Forwarding Ratio (PFR) which can be calculated as below

$$PFR = \frac{\text{No. of packets forwarded}}{\text{No. of packets received}} \tag{2}$$

X_2 - *Link Alive Time (LAT).* This metric is used to find the approximate lifetime of a given wireless link using the mobility and location information of nodes. Link Alive Time between two nodes can be estimated using the information such as current position of the nodes, their direction of movement, and their transmission ranges. The wireless link between nodes a and b with transmission range T_X, which are moving at velocity V_a and V_b at angles T_a and T_b can be estimated as below

$$LAT = \frac{-(pq+rs) + (p^2+r^2)Tx^2 - (ps-qr)^2}{p2+q2} \tag{3}$$

Where

$P = V_a\cos T_a - V_b\cos T_b$

$Q = X_a - X_b$

$R = V_a\sin T_a - V_b\sin T_b$

$S = Y_a - Y_b$

X_3 - *Average Throughput.* Throughput is defined as the total amount of data a receiver actually receives from the sender S divided by the time it takes for R to get the last packet. The average throughput in Kbps can be computed as shown below:

$$\text{Average throughput} = (X * L)/t \tag{4}$$

Where *(X) denotes the number of the packets successfully received, (L) is the packet size and (t) is the simulation time.* The node which improves the throughput will be selected for routing.

X_4 - *Node Buffer Size.* Every node need to store the data packets for particular time periods to recover it in case of any route failures. Each intermediate node should store the selected feasible nodes temporarily for future routing. So the node with highest buffer size will be given priority.

X_5 - *Node Mobility.* In MANET the network topology is highly dynamic due to the movement of nodes; hence routing suffers from frequent path breaks. Mobility of a node can be measured using the following the formula

$$M_v = \frac{1}{T}\sum_{t=1}^{T} \sqrt{(X_t - X_{t-1})^2 + (Y_t - Y_{t-1})^2} \tag{5}$$

where (X_t, Y_t) and (X_{t-1}, Y_{t-1}) are the coordinates of a node at time t and t−1. The current time is T. So the node with highest mobility will be given lower weight.

X_6 - *Load on Node.* Data packets will be buffered in the nodes when the routes are busy. Number of packets waited in the buffer queue for transmission is to be calculated in order to choose the node with minimum load.

X_7 - *Energy.* In MANET power consumption by the nodes is a serious factor to be taken into consideration by routing protocols. The energy efficiency of a node is defined as the ratio of the amount of data delivered by the node to the total energy expended or using loss power using the distance.

The remaining energy of the nodes can be calculated using

$$E_R = E_T - t * d(u,v)^n \tag{6}$$

Where

E_R	– Remaining energy
E_T	– Total energy
t	– a constant
n	– The path loss exponent indicating the loss power with distance from the transmitter
d(u, v)	– distance between two nodes u and v

Request Zone Creation Phase:

Step 1: If (X_s, Y_s) location of S, the location of D at time t_0 is (X_d, Y_d).

Step 2: If destination is in moving with the nodal speed v, the distance travelled at time t_1 is

$$dis = v * (t_1 - t_0). \tag{7}$$

Step 3: Then location of destination D may be at any one of the following locations.

$$
\begin{aligned}
&\text{(i) } (X_{d+dis}, Y_d \\
&\text{(ii) } (X_d, Y_{d+dis}) \\
&\text{(iii) } (X_{d-dis}, Y_d) \\
&\text{(iv) } (X_d, Y_{d-dis})
\end{aligned}
\tag{8}
$$

Step 4: (X_{id}, Y_{id}) is the position of D at time ti.

Step 5: A request zone is created between (Xs, Ys) and (X_{id}, Y_{id}) as shown in Fig. 2

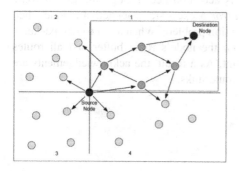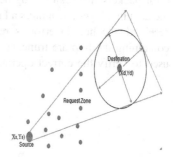

Fig. 2. Request zone creation

Route Discovery

Step 1: If a source node S intends to send data to destination D. First S has to find whether the route for D is available in the buffer or not.

Step 2: If there is a route then start communication.

Step 3: Otherwise starts route discovery.

Step 4: A request zone is created based on the location of S and D.

Step 5: Only feasible neighbor nodes are selected for routing if $(X_{id} - X_s) > 0$ then intermediate nodes at $(X_k\ Y_k)$ are feasible only when $(X_s <= X_k <= X_{id})$ else nodes in the range $(X_{id} <= X_k <= X_s)$ are feasible. Likewise y coordinate constrains are considered $(Y_s <= Y_k <= Y_{id})$ or $(Y_{id} <= Y_k <= Y_s)$

Step 6: Among the feasible nodes three nodes with highest GP value is considered and stored in buffer of S.

Step 7: The S send RouteRequest *packet to* one optimal node among these three nodes which satisfies MAX $\sum W_i X_i$

Step 8: If the source is X the successive nodes may be named as X1, X2 and X3. Suppose X1 has picked, it consists of three nodes and that is indicated as X1Y1, X1Y2, X1Y3 and so on.

Step 9: Repeat these procedure until destination reached.

Step 10: Destination will send **Route Reply.** After receiving **Route Reply** source starts communication.

Step 11: If any failure in the route, the source will select next optimal node which is in the buffer for data communication. (Fig. 3)

Route Maintenance: Route maintenance is accomplished through the use of route error packets and acknowledgment. *Route error* packets are generated at a node when the data link layer encounters a fatal transmission problem. When a route error packet is received, the hop in error is removed from the node's route buffer and all routes containing the hop are truncated at that point. As a result, the acknowledgements are used to verify the correct operation of the route links.

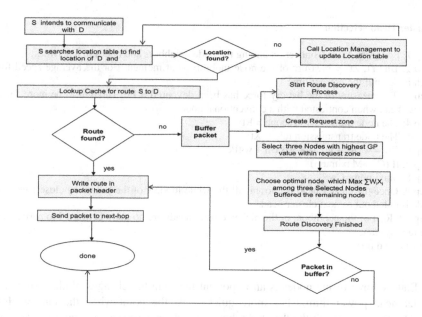

Fig. 3. DTV algorithm

3.2 Prophecy Weighted Clustering Algorithm

Clustering refers to a technique in which MANET is divided into different virtual group. Generally nodes which are geographically adjacent are allocated into the same cluster. Clusters are driven by set of protocols based on behaviors and characteristics of the node. This enables the network to become manageable. Various clustering techniques [14–17] allow fast connection and also better routing and topology management of MANET. Clustering technique ensure scalability and load balancing in MANETs, increases the system capacity by facilitating the spatial reuse of resources and enhances the co-ordination of transmission activities by electing central authority. To provide a stable and smaller vision of ad hoc network clustered structure can be created within request zone of DTV. This algorithm is composed of two phases: *First Phase*: Clusters are formed and an appropriate Cluster Head (CH) is elected for cluster. *Second Phase:* To overcome the disadvantage of the single cluster, the multiple cluster head are elected and these are maintained by using Cluster Head Probability.

Cluster Head selection

Step 1: For every Node Connection among the node neighbors and nodes in a network;
Step 2: Bn.(Yt, Zt) = measure of the node mobility w at moment t or the average speed for each node until the present time T;
Step 3: The measure of the battery force has been devoured = Pc. Thought to be more for a cluster head when contrasted with a conventional node;
Step 4: The node transmission scope = Rt;
Step 5: The node transmission rate = Xt;
Step 6: Cw = w1Rt + w2Xt + w3 Bn + w4Pc
Step 7: If (Cw>Maximum)
Step 8: Maximum=Cw;
Step 9: Once the cluster head is chosen, all the while in view of the following closest weight node is doled out as next cluster head.
Step 10: Rest of the neighbors of the picked cluster heads are no more permitted to partake in the race system;
Step 11: end for;

Battery force of the nodes is an important factor in the changes of clustering. The cluster head power diminishes more quickly. At the point when the cluster Head Probability falls beneath the threshold then the node is never again having the capacity to execute as a cluster head and another cluster head is picked.

3.3   Security Mechanism

Unlike the traditional wired Internet, where dedicated routers controlled by the Internet Service Providers (ISPs) exist, in ad hoc wireless networks, nodes act both as regular terminals and also as routers for other nodes. In the absence of dedicated routers, providing security becomes a challenging task in these networks [19, 20]. Various other factors which make the task of ensuring secure communication in ad hoc wireless networks difficult include the mobility of nodes, a promiscuous mode of operation, limited processing power, and limited availability of resources such as battery power, bandwidth, and memory. To provide secured routing a security mechanism is integrated in the proposed DTVOSSP. The Cluster Head (CH) distributes a randomly generated key to authenticate the cluster members at regular interval. Only authenticated nodes selected for data transmission. Every *Route Reply* will contain Time To Reply (TTR), and update hop count at every hop. In order to find out the cooperative nodes the trust value of each intermediate node compared with threshold value. Trust value of every node can be computed based on Trace and Hope Based Method (THBM) in which the CH premeditate the trust value through tracing all member nodes and Report Based Method (RBM) [21] in which the trust value is reported by node to its neighbor node. If all conditions satisfied then Secured Data communication starts (Fig. 4) [18, 22, 23].

Report Based Method &Trace and Hope Based Method (RBM & THBM)

Step 1: Collect data for Rp, Sp, f, d, m, i.

Step 2: Find the threshold values associated to each behavior fn, dn, mn.

Step 3: Calculate ratio fs, ds, ms, is of each behavior and Rp, Sp total sent or received packet Accordingly.

Step 4: Calculate the deviation fd, dd, md, id from the corresponding threshold.

fs = f / Rp and fd= fn / fs

ds = d / Rp and dd = dn/ ds

ms = m / Rp and md= mn/ ms

is = i / Sp and id= in/ is

Step 5: Calculate the corresponding direct trust value using the formula

Trust (t) = (w1*fd) - (w2*dd) + (w3*md) + (w4*id)

Trust Metric

f = No. of packets forwarded

d = No. of packets dropped

m = No. of packets misrouted

i = No. of packets falsely injected

Rp= Total no. of packets received by B sent from A

Sp= Total no. of packets sent by B to A

Trust value =(Trust value of RBM + Trust value of THBM)/2

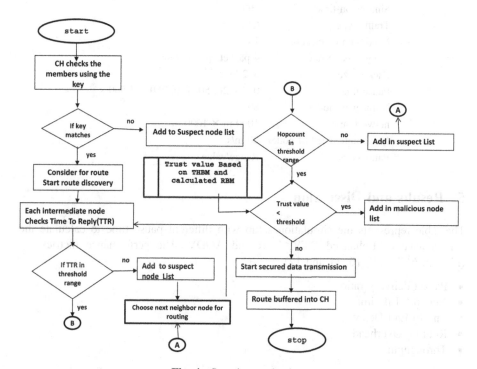

Fig. 4. Security mechanism

3.4 Features of DTVOSSP

It reduces route request flood and delay in route discovery. The protocol minimizes the number of nodes for Route computation and maintenance. It optimally uses scarce resources such as bandwidth, computing power, memory and battery power. It is an adaptive to frequent topology changes caused by mobility of nodes. It provides an efficient route and a certain level of Quality of Service as well as ensures scalability and load balancing in MANETs. Increase of system capacity by facilitating the spatial reuse of resources. Enhances the co-ordination of transmission activities by electing central authority and provides a stable and smaller vision of ad hoc network.

4 Experimental Setup

With the network simulator, DTV model is employed for simulation with 80 mobile nodes in the 1000 m × 1000 m for the simulation time of 500 s. In the network, the node transmission will be 250 m for all nodes. For traffic source, the CBR (Constant Bit Rate) is used Table 1.

Table 1. Experimental parameters and setup

Parameters	Values
Routing protocols	Enhanced DTV, AODV and DSR
Simulation time	500 s
Traffic type	CBR
Maximum connections	10
Transmission rate	4 packets per second
Packet size	512 byte
Pause time	0, 10, 20, 50, 100, 250 and 500 s
Number of nodes	80
network area	1000 m × 1000 m
Maximum speed of nodes	20 m/s
Parameters	Range/value

5 Results and Discussion

The table represents the simulation setup with different pause time to calculate the performance of Enhanced DTV, DSR and AODV. The performance metrics are (Figs. 5, 6, 7, 8, 9, 10, 11 and 12)

- Packet delivery ratio
- Network Lifetime
- End to End Delay
- Routing Overhead
- Throughput

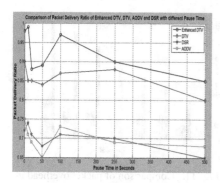

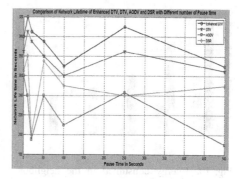

Fig. 5. Packet delivery ratio

Fig. 6. Network lifetime

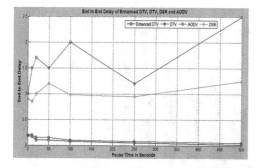

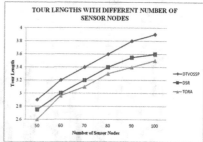

Fig. 7. Comparision of end-to-end delay

Fig. 8. Comparision of throughputs

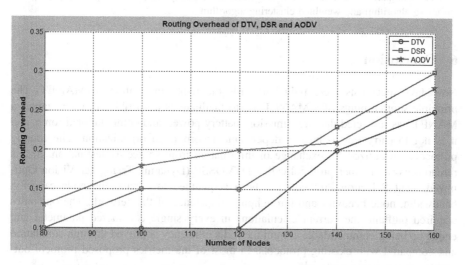

Fig. 9. Routing overhead comparison between DTV, DSR and AODV

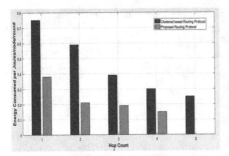 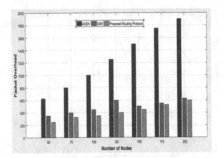

Fig. 10. Energy consumed in the neighbor discovery DSR and proposed routing protocol

Fig. 11. Comparison of packet overhead in AODV, DSR and proposed routing protocol

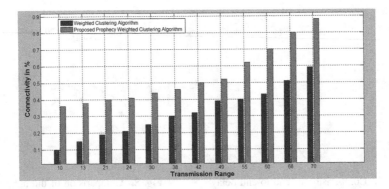

Fig. 12. Comparison of transmission range and connectivity for proposed prophecy weighted clustering algorithm and weighted clustering algorithm

6 Conclusion

Nowadays specialists were pulled towards the rising innovation of MANET. The principle shortcomings of a MANET are to facilitate as asset obliged, for instance, a MANET has restricted data transmission, battery power, and computational force, and it is not to carry out to have a dependable incorporated organization. The routing protocols are utilized for exchange of information from source to destination. In this research work, another protocol called DTVOSSP (Dynamic Triangular Vision Optimized Slant Selection Protocol) has been acquainted to expand the node capacity, bandwidth, node behavior and throughput competence of the network. The outcomes acquired outflank the current calculations in every single considered parameter and ended up being commendable commitment. Its efficiency is proved through simulation and compared with existing protocols. In most of the metrics proposed protocol outperform the existing protocols.

References

1. Navitha, S., Velmurugan, T.: A survey on the simulation models and results of routing protocols in mobile ad-hoc networks. Int. J. Commun. Netw. Syst. **04**(02), 10–15 (2015)
2. Singh, S.K., Duvvuru, R., Singh, J.P.: TCP and UDP based performance evaluation of proactive and reactive routing protocols using mobility models in MANETS. Int. J. Inf. Commun. Technol. **7**(6), 632–644 (2015)
3. Maakar, S.K., Singh, Y.: Traffic pattern based performance comparison of two proactive MANET routing protocols using Manhattan grid mobility model. Int. J. Comput. Appl. **114** (14), 26–31 (2015)
4. Bansal, B., Tripathy, M.R., Goyal, D., Goyal, M.: Improved routing protocol for MANET. In: 2015 5th International Conference on Advanced Computing and Communication Technologies (ACCT), pp. 340–346, 21–22 February 2015
5. Perkins, C.E., Bhagwat, P.: Highly dynamic destination-sequenced distance-vector routing (DSDV) for mobile computers. In: Proceedings of ACM SIGCOMM (1994)
6. Johnson, D.B., Maltz, D.A.: Dynamic source routing in ad hoc networks. In: Imielinski, T., Korth, H. (eds.) Mobile Computing, pp. 153–181. Kluwer Academic Publishers, Dordrecht (1996)
7. Johnson, D.B., Maltz, D.A., Hu, Y.: Dynamic SourceRouting Protocol for Mobile Ad Hoc Networks (DSR), July 2004. http://www.ietf.org/internet-drafts/draft-ietf-manet-dsr-10.txt
8. Perkins, C.E., Royer, E.M.: Ad hoc on-demand distance vector routing. In: Proceedings of IEEE Workshop on Mobile Computing Systems and Applications, pp. 90–100 (1999)
9. Park, V., Corson, S.: Temporary-ordered routing algorithm (TORA). Internet Draft: draft-ietf-manettora-spec-04.txt (2001)
10. Gupta, A.K., Sadawarti, H., Verma, A.K.: Performance analysis of AODV, DSR & TORA routing protocols. IACSIT Int. J. Eng. Technol. **2**(2), 226–231 (2010)
11. Ko, Y.-B., Vaidya, N.H.: Location-aided routing (LAR) in mobile ad hoc networks. Wirel. Netw. **6**, 307–321 (2000)
12. Sharma, R., Lobiyal, D.K.: Proficiency analysis of AODV, DSR and TORA Ad-hoc routing protocols for energy holes problem in wireless sensor networks. In: 3rd International Conference on Recent Trends in Computing 2015 (ICRTC-2015), pp. 1057–1066 (2015)
13. Yi, J., Fuerters, J., Clausen, T.: Jitter Consideration for Reactive Protocols in Mobile Ad Hoc Networks. Network Working Group, 13 July 2013
14. Naser, M.A.: Study analysis and propose enhancement on the performance of cluster routing algorithm in a mobile ad hoc network. J. Adv. Comput. Sci. Technol. Res. **4**(1), 12–21 (2014)
15. Bagwar, A., Joshi, P., Rathi, V., Soni, V.: Routing protocol behaviour with multiple cluster head gateway in mobile ad hoc network. Int. J. Eng. Sci. Technol. (IJEST) **3**(7), 133 (2011). ISSN 0975-5462
16. Agrawal, V., Chauhan, H.: An overview of security issues in mobile ad hoc networks. Int. J. Comput. Eng. Sci. **1**, 1–8 (2014)
17. Chatterjee, M., Das, S.K., Turgut, D.: A weight based distributed clustering algorithm for mobile ad hoc networks. In: Valero, M., Prasanna, Viktor, K., Vajapeyam, S. (eds.) HiPC 2000. LNCS, vol. 1970, pp. 511–521. Springer, Heidelberg (2000). doi:10.1007/3-540-44467-X_47
18. Chatterjee, M., Das, S.K., Turgut, D.: WCA: a weighted clustering algorithm for mobile ad hoc networks. Cluster Comput. **5**, 193–204 (2002). Kluwer Academic Publishers, The Netherlands

19. Anandhi, M., Ravi, T.N.: An appraisal of attacks in MANET and its fortification methods. Int. J. Adv. Res. Comput. Sci. (IJARCS) **5**, 216–220 (2014)

20. Anandhi,M., Ravi, T.N.: Scalable, power aware and efficient cluster for secured MANET. Int. J. Comput. Sci. Inf. Secur. (IJCSIS) **13**, 79–82 (2015). ISSN 1947-5500 USA, Impact factor 0.476

21. Anandhi, M., Ravi, T.N.: A novel framework for prevent the denial of service attacks in MANET. Int. J. Recent Innov. Trends Comput. Commun. (IJRITCC), Sci. J. **3**, 3201–3206 (2015). ISSN 2321-8169

22. Anandhi, M., Ravi, T.N.: Dynamic triangular vision and optimized slant selection protocol for mobile ad-hoc network. SCOPUS Index. Int. J. Appl. Eng. Res. (IJAER) **10**(82), 508–514 (2015). Print ISSN 0973-4562, Online ISSN 1087-1090 (Impact factor: 1.8233)

23. Anandhi, M., Ravi, T.N.: RAN-DTVOSSP routing algorithm for mobile adhoc networks. Middle-East J. Sci. Res. **24**, 199–207 (2016). ISSN 1990-9233

Community Detection Based on Girvan Newman Algorithm and Link Analysis of Social Media

K. Sathiyakumari[✉] and M.S. Vijaya

PSGR Krishnammal College for Women, Avinashi Road, Peelamedu,
Coimbatore 641004, Tamilnadu, India
{sathiyakumari,msvijaya}@psgrkc.com

Abstract. Social networks have acquired much attention recently, largely due to the success of online social networking sites and media sharing sites. In such networks, rigorous and complex interactions occur among numerous one-of-a-kind entities, main to massive statistics networks with notable enterprise capacity. Community detection is an unsupervised learning task that determines the community groups based on common interests, occupation, modules and their hierarchical organization, using the information encoded in the graph topology. Finding communities from the social network is a difficult task because of its topology and overlapping of different communities. In this research, the Girvan-Newman algorithm based on Edge-Betweenness Modularity and Link Analysis (EBMLA) is used for detecting communities in networks with node attributes. The twitter data of the well-known cricket player is used right here and community of friends and fans is analyzed based on three exclusive centrality measures together with a degree, betweenness, and closeness centrality. Also, the strength of extracted communities is evaluated based on modularity score using proposed method and the experiment results confirmed that the cricket player's network is dense.

Keywords: Edge-Betweenness · Modularity · Degree · Closeness · Community detection · Social network

1 Introduction

The developing use of the internet has brought about the development of networked interplay environments inclusive of social networks. Social networks are graph structures whose nodes represent people, organizations or other entities, and whose edges represent relationship, interaction, collaboration, or influence between entities. The rims in the network connecting the entities might also have a path indicating the drift from one entity to the other; and a power denoting how plenty, how often, or how critical the connection is. Researchers are increasingly interested in addressing a wide range of challenges exist in these social network systems.

In recent years, social community research has been completed the use of large amount of data collected from on-line interactions and from explicit courting hyperlinks in online social community systems including facebook, Twitter, LinkedIn,

© Springer Nature Singapore Pte Ltd. 2016
S. Subramanian et al. (Eds.): CSI 2016, CCIS 679, pp. 223–234, 2016.
DOI: 10.1007/978-981-10-3274-5_18

Flickr, instant Messenger, and so on. Twitter is pretty rated as a new shape of media and utilized in numerous fields, consisting of corporate advertising and marketing, education, broadcasting and etc. Structural characteristics of such social networks can be explored using socio metrics to understand the structure of the network, the properties of links, the roles of entities, information flows, evolution of networks, clusters/communities in a network, nodes in a cluster, center node of the cluster/network, and nodes on the periphery etc. To discover functionally associated objects from network groups [1, 2] allow us to observe interaction modules, lacking characteristic values and are expecting unobserved connections among nodes [3]. The nodes have many relationships among themselves in communities to percentage common homes or attributes. The identifying community is a trouble of clustering nodes into small communities and a node may be belonging to a couple of communities straight away in a community structure.

Two exclusive assets of facts are used to carry out the clustering mission, first is ready nodes and its attributes and the second one is ready the connection between nodes. The attributes of nodes in community structure are known properties of users like network profile, author publication, publication histories which helps to determines similar nodes and community module to which the node belongs. The connection between nodes provides information about friendships, authors collaborate, followers, and topic interactions.

A few clustering algorithms [4, 5] employ node attributes but ignores the relationships among nodes. However, the network detection algorithms use corporations of nodes which can be densely linked [6, 7] but ignore the node attributes. By means of the use of those two sources of data, the positive algorithm fails to describe the critical shape in a community. For example, attributes may also inform about which community node with few hyperlinks belonging to and it is difficult to determine from community shape alone. On the contrary, the community offers detail about nodes belongs to the equal community even someone of the nodes has no attributes values. Node attributes can balance the network shape which ends up in an extra correct detection of communities. Thus community detection becomes challenging task when taking into account of both nodes attributes and network topology.

The proposed method overcomes the above hassle by identifying groups based totally on the node and its attributes with the aid of implementing Girvan-Newman set of rules.

2 Related Work

A community is a densely linked subset of nodes that is sparsely connected to the remaining network. Social networks are a combination of vital heterogeneities in complex networks, which includes collaboration networks and interaction networks. Online social networking applications are used to represent and model the social ties among people. Finding groups inside an arbitrary community may be a computationally hard task. Various research dealings in recent past years have been conducted on the topic of community detection and some of the important research works are mention below.

Chen and Yuan have referred to that counting all feasible shortest paths in the calculation of the brink betweenness can make unbalanced partitions, with groups of very distinctive length, and proposed to rely on handiest non-redundant paths, i.e. paths whose endpoints are all special from every different. The resulting betweenness confirmed higher consequences than general facet betweenness for blended clusters at the benchmark graphs of Girvan and Newman. Holme et al. have used a changed model of the algorithm wherein vertices, instead of edges, have been removed. A centrality measure for the vertices, proportional to their web page betweenness that became inversely proportional to their in-degree turned into selected to perceive boundary vertices, which have been then iteratively removed with all their edges. Only the in-degree of a vertex becomes used as it indicated the number of substrates to a metabolic reaction concerning that vertex [8].

One of the most popular algorithms changed into provided via Newman and Girvan (denoted GN) [9, 10] which turned into a divisive hierarchical clustering set of rules. Edge removal divided network to groups, the rims to take away had been chosen through the usage of betweenness measure. The concept changed into that if companies are related by a few edges between them, then all the paths between vertices in a single group to vertices in different companies blanketed these edges. Paths give ratings to edges betweenness, with the aid of accounting all the paths passing via each aspect and removing the threshold with the maximal rating, hyperlinks inside the community had been broken. This system was repeated and turned into divided into smaller paths until a stop criterion is reached, this criterion become modularity. A hybrid model of this method in [11] and a faster version primarily based on the equal strategy in [12] become proposed.

Approaches to network detection based totally on the genetic set of rules have been available in [13–15]. A genetic method proposed by using [16] applied an algorithm that used a health characteristic which recognized businesses of vertices within the network that have dense intra connections and sparser inter connections.

In [17, 18] authors proposed a genetic set of rules that uses Newman and Girvan fitness feature for measuring network modularity. Characters become covered of N genes that N changed into the node range. The ith gene corresponds to a jth node, and its fee becomes the identifier of node i. Authors used a non-fashionable one-manner crossover in which, given two people A and B, a network identifier j was selected randomly, and the identifier j of nodes $j_1,...,j_h$ of A become transferred to the identical nodes of B.

On this research, the Girvan-Newman set of rules based totally on part-Betweenness Modularity and link analysis (EBMLA) is applied for discovering groups in networks with node attributes. The twitter data of the famous cricket participant is taken for have a look at and network of friends and followers is analyzed based totally on 3 different centrality measures along with the degree, betweenness, closeness centrality, and modularity score.

3 Girvan-Newman Algorithm

3.1 Girvan-Newman Algorithm Based on Edge-Betweenness Modularity and Link Analysis

The Girvan and Newman is a general community finding algorithm. It performs natural divisions among the vertices without requiring the researcher to specify the numbers of communities are present, or placing limitations on their sizes, and without showing the pathologies evident in the hierarchical clustering methods. Girvan and Newman [19] have proposed an algorithm which has three definitive features (1) edges are gradually removed from a network (2) the edges to be removed are chosen by computing betweenness scores (3) the betweenness scores are recomputed for removal of each edge.

As a degree of traffic flows Girvan and Newman use part betweenness a generalization to the edge of the renowned vertex betweenness of Freeman [20, 21]. The betweenness of an edge is defined because of the quantity of shortest paths among vertex pairs. This quantity can be calculated for all edges in the time complexity of O (mn) on a graph with m edges and n vertices [22, 23].

Newman and Girvan [23] define a degree called modularity, which is a numerical index that shows proper separation among nodes. For a separation with g organizations, define as $g \times g$ matrix e. Whose thing e_{ij} is the fraction of edges in the authentic network that connects vertices in institution i to those in institution j. Then the modularity is described as

$$Q = \sum_i e_{ii} - \sum_{ijk} e_{ij}e_{ki} = \text{Tr } e - \left\| e^2 \right\|,$$

wherein shows the sum of all elements of x, Q is the fraction of all edges that lie within groups minus the predictable value of the same amount in a graph in which the vertices have the same tiers however edges are positioned at random with outlook upon the communities. The Q = 0 indicates that community shape isn't any more potent than could be expected by using random chance and values other than 0 represent deviations from randomness. Restricted peaks inside the modularity for the duration of the progress of the community shape set of rules imply correct divisions of the community.

3.2 Girvan-Newman Partitioning Algorithm

Successively Deleting Edges of High-Betweenness

Step 1: discover the threshold or multiple edges with the best betweenness, if there may be a tie in betweenness then eliminate the rims from the graph. This technique may additionally spill the graph into numerous components; it makes first level partition of the graph.

Step 2: Recalculate all betweenness values and then remove the edges/edge with high betweenness value. Again split the first level region into several components such that there are nested within larger regions of the graph.

Step 3: Repeat steps (1) and (2) until edges continue to be within graph.

Computing Betweenness Values
For each node A:

Step 1: Do Breath First seek to start at node A
Step 2: remember the quantity of shortest paths from A to every different node
Step 3: determine the quantity of flow from A to all other nodes.

3.3 Centrality Measures and Modularity Scores

Centrality measures are used to discover the node's relative significance inside groups by using summarizing structural relation with different nodes. The three simple centrality measure targeted on this work are a degree, closeness, and betweenness.

Degree. The degree centrality represents the wide variety of connections a selected node has. In a directed graph, wherein the route of the node is applicable, there may be a differentiation between the in-degree and out-degree; the quantity of hyperlinks a specific node receives is in-diploma, and the range of links a selected node sends is out-degree. The sum of in-degree and the out-degree offers the degree measure. The following method gives degree and normalized degree centrality ratings.

Degree centrality

$$C_D(v) = deg(v)$$

Normalized degree centrality

$$C_D(v) = deg(v)/g - 1$$

where g is the size of the group.

Closeness. The closeness measure represents imply of the geodesic distances among a particular node with other nodes related to it. It is a measure of ways long a message will take to unfold during the network from a specific node n sequentially. It also describes the speed of the message within social systems. Closeness is primarily based on the period of the average shortest direction among a vertex and all of the vertices in the graph. The subsequent formulation is used to calculate closeness centrality.

Closeness Centrality

$$C_c(n_i) = \left[\sum_{j=1}^{g} d(n_i, n_j) \right]^{-1}$$

Normalized Closeness Centrality

$$C'_c(n_i) = (C_c(n_i))(g - 1)$$

Betweenness. The betweenness measure quantifies the quantity of times a node acts as a bridge alongside the shortest direction among other nodes. It's miles a measure for quantifying the manipulate of the node at the conversation between nodes in a social network. It also represents how a long way a message can reach inside a network from a specific node 'n' and additionally describes the span of the message within social systems. Nodes that arise on many shortest paths among other nodes have higher betweenness than those that do not. This is vertices that have an excessive possibility to occur on a randomly chosen shortest direction among two randomly chosen vertices have an excessive betweenness. The following formulas are used to determine betweenness centrality measure.

Betweenness Centrality

$$C_B(n_i) = \sum_{j<k} g_{jk}(n_i)/g_{jk}$$

where g_{jk} = the number of geodesics connecting jk, and $g_{jk}(ni)$ = the number that actor i is on.

Normalized betweenness centrality measure

$$C'_B(n_i) = C_B(n_i)/[(g-1)(g-2)/2]$$

Modularity. The modularity Q is proposed via Newman and Girvan [23] as a degree of the nice of a selected division of a network, and is defined as follows:

Q = (range of edges inside communities) − (predicted wide variety of such edges)

The modularity Q measures the fraction of the edges within the community that join vertices of the same type, i.e., inside-community edges, minus the expected value of the same quantity in a community with the equal network department however with random connections among the vertices. If the variety of inside community edges is not any higher than random, Q = zero. A price of Q this is near 1, which is the maximum, indicates strong community shape. Q usually falls inside the range from 0.3 to 0.7 and excessive values are rare.

4   Experiments and Results

The proposed framework includes four phases: twitter facts, directed network, Girvan-Newman algorithm, and modularity score. Every phase is described in following sections and the architecture of the proposed system is shown in Fig. 1.

Real time twitter statistics was extracted from twitter API 1.1 the use of R 3.3.1 tool. A directed network is created using Twitter buddies/followers listing as the graph. In this directed community, three centrality measures degree, closeness, and betweenness are used for the stage of evaluation of network Girvan-Newman algorithm

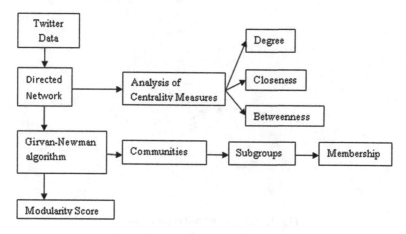

Fig. 1. Community detection framework

is used to detect communities and subgroups. The size of subgroups is found using Girvan-Newman algorithm of this network. The algorithm also detects modularity score of the community.

Girvan-Newman algorithm is implemented for community detection based on edge betweenness. Analysis of the social network is carried out using various centrality measures such as degree, closeness, and betweenness. These centrality measures are evaluated with various properties like minimum and maximum values of in-degree, out-degree, in-closeness, out-closeness, and betweenness. The real-time data is collected using the twitter application programming interface 1.1 for this research work. Nine thousands records of friends and followers list of the famous cricket player have been crawl from his twitter account. The data is collected at run time from twitter network using R3.3.1, a statistical tool.

Figure 2 shows the cricket player's initial community network and Fig. 3 depicts the relationship types of community network such as friends and followers both friends and followers of the initial network. This network has 7095 edges and 6831 vertices.

Degree, closeness, and betweenness are the three centrality measures that are evaluated for the above network using R script. The number of connections that a

Fig. 2. Cricket player's initial network

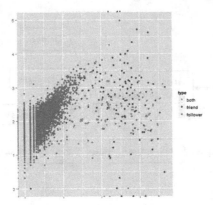

Fig. 3. Friends and followers network

particular node makes is called the degree centrality. The Twitter network is a directed graph and a node encompasses both in-degree and out-degree. The number of arcs from a node to other nodes is out-degree and it is 95 for this network. The in-degree is the number of arcs coming into a node from other nodes and it is 7000 on the same network. The total degree centrality measure is 7095. The histogram representation of in-degree, out-degree and total degree measurement for the cricket player's network is shown in Figs. 4 and 5. The minimum and maximum values of in-degree and out-degree measures are given in Table 1.

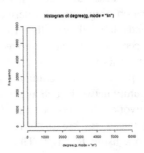

Fig. 4. In-degree of given network

Similarly, the closeness centrality measure is evaluated for the same directed graph. The closeness measure represents the shortest path between nodes connected with it. The out-closeness is 0.0000000217 for this network. The in-closeness is 0.00000161 for the same network. The total closeness centrality is 0.0000000217. Figures 6 and 7 displays the histogram representation of in-closeness and out-closeness of this network. The minimum and maximum values of in-closeness and out-closeness measures are given in Table 1.

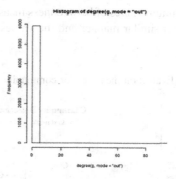

Fig. 5. Out-degree of given network

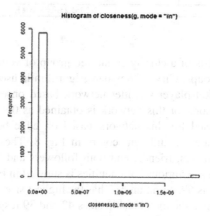

Fig. 6. In-closeness of given network

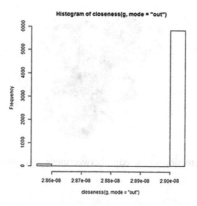

Fig. 7. Out-closeness of given network

The minimum and maximum values of betweenness measures are computed for this cricket player's network in a similar manner and the values are given in Table 1.

Table 1. Evaluation measures for community detection

Measures / Limitations	Degree			Community Detection Measures Closeness			Betweenness
	In	Out	Total degree	In	Out	Total Closeness	
Min	1	1	1	0.0000000214	0.0000000214	0.0000000214	1
Max	7000	95	7095	0.00000161	0.0000000217	0.0000000217	640295
						Modularity Score is : 0.91	

A community consists of a closely connected group of vertices, with only meager connections to other groups. Girvan-Newman Algorithm is used here to find different communities from cricket player's twitter network based on edge betweenness measure. The modularity score for this network is obtained as 0.91. Thirty-nine different communities are extracted for this network based on edge betweenness modularity measure and demonstrated in different colors in Fig. 8. These 39 communities are clustered based on followers, friends, and both followers and friends in the network. The distribution of nodes in various communities is showed in Fig. 9. The membership of size of Community 1 is 69, community 2 has the highest size with 166 memberships. Communities 3 and 4 have the membership sizes 42 and 39 respectively. Communities 5 and 7 have the same membership size 37 and so on.

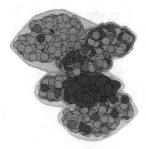

Fig. 8. Community detection using edge betweenness algorithm (Color figure online)

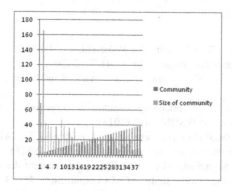

Fig. 9. Community size

5 Discussion and Findings

In this research work, the out-degree is 95 and in-degree are 7000 for the cricket player's network. An entity or node is an active player, hub when it has high degree centrality and obtains an advantaged position in the network. Since closeness centrality is low, the node has slow interaction to other entities in a network. The node has a better influence over the other nodes in the network and is in a powerful position because the betweenness centrality is 468 for this network. The modularity score obtained through Edge-Betweenness and Link algorithm is 0.91; it is proved that the cricket player's friends and followers network are highly dense. Also, the Girvan-Newman algorithm has detected 39 different communities from the cricket player's network and found that out of 39 communities, 5 communities are dense.

6 Conclusion and Future Work

This work elucidates the application of Girvan-Newman algorithm based on Edge-Betweenness and Link Analysis for detecting communities from networks with node attributes and its properties. The real time twitter directed network of a cricket player is used to carry out social network analysis with various centrality measures. The modularity value 0.91 of the tested network by EBMLA confirms that the network is dense and the algorithm is efficient in finding different communities. As a scope for further work more network properties can be used for social network analysis and more interpretation can be drawn. Also, other community detection algorithms can be adopted for detecting communities and finding significant node attributes for each community.

References

1. Coscia, M., Rossetti, G., Giannotti, F., Pedreschi, D.: DEMON: a local-first discovery method for overlapping communities. In: KDD 2012 (2012)
2. Girvan, M., Newman, M.: Community structure in social and biological networks. PNAS **99**, 7821–7826 (2002)
3. Yang, J., Leskovec, J.: Overlapping community detection at scale: a non-negative factorization approach. In: WSDM 2013 (2013)
4. Johnson, S.: Hierarchical clustering schemes. Psychometrika **32**, 241–254 (1967)
5. Fortunato, S.: Community detection in graphs. Phys. Rep. **486**, 75–174 (2010)
6. Xie, J., Kelley, S., Szymanski, B.K.: Overlapping community detection in networks: the state of the art and comparative study. ACM Comput. Surv. **45**, 43 (2013)
7. Fortunato, S.: Community detection in graphs. J. Phys. Rep. **486**(3–5), 75–147 (2010)
8. Newman, M.E.J., Girvan, M.: Finding and evaluating community structure in networks. J. Phys. Rev. E **69**(2), 026113 (2004)
9. Girvan, M., Newman, M.E.J.: Community structure in social and biological networks. In: Proceedings of the National Academy of Science, USA, pp. 7821–7826 (2002)
10. Newman, M.E.J.: Fast algorithm for detecting community structure in networks. J. Phys. Rev. E **69**(6), 066133 (2004)
11. Clauset, A., Newman, M.E.J., Moore, C.: Finding community structure in very large networks. J. Phys. Rev. E **70**(6), 066111 (2004)
12. Nandini, R.U., Reka, A., Soundar, K.: Near linear time algorithm to detect community structures in large-scale networks. J. Phys. Rev. E **76**(3), 036106 (2007)
13. Guardiola, X., Guimera, R., Arenas, A., Guilera, A.D., Antonio, L.: Macro- and micro-structure of trust networks, pp. 1–5 (2002). arXiv:cond-mat
14. Narasimhamurthy, A., Greene, D., Hurley, N., Cunningham, P.: Scaling community finding algorithms to work for large networks through problem decomposition. In: 19th Irish Conference on Artificial Intelligence and Cognitive Science (AICS 2008), Cork, Ireland (2008)
15. Pizzuti, C.: Community detection in social networks with genetic algorithms. In: Proceedings of the 10th Annual Conference on Genetic and Evolutionary Computation, pp. 1137–1138 (2008)
16. Tasgin, M., Bingol, H.: Communities detection in complex networks using genetic algorithm. In: Proceeding of the European Conference on Complex Systems (ECSS), UK, pp. 1–6 (2006)
17. Tasgin, M., Herdagdelen, A., Bingol, H.: Communities detection in complex networks using genetic algorithms. Community detection in complex networks using genetic algorithms. arXiv preprint arXiv:0711.0491, pp. 1–6 (2007)
18. Girvan, M., Newman, M.E.J.: Community structure in social and biological networks. Proc. Natl. Acad. Sci. U.S.A. **99**, 7821–7826 (2002)
19. Freeman, L.C.: A set of measures of centrality based upon betweenness. Sociometry **40**, 35–41 (1977)
20. Anthonisse, J.M.: The rush in a directed graph. Technical report BN9/71, Stichting Mathematicsh Centrum, Amsterdam (1971)
21. Newman, M.E.J.: Scientific collaboration networks: II. Shortest paths, weighted networks, and centrality. Phys. Rev. **E64**, 016132 (2001)
22. Brandes, U.: A faster algorithm for betweenness centrality. J. Math. Sociol. **25**, 163–177 (2001)
23. Newman, M.E.J., Girvan, M.: Finding and evaluating community structure in networks (2003). Preprint arXiv:cond-mat/0308217

IT for Society

GIS Based Smart Energy Infrastructure Architecture and Revenue Administration

Shailesh Kumar Shrivastava[1]([⊠]), S.K. Mahendran[2],
and Amar Nath Pandey[3]

[1] National Informatics Centre, Government of India,
Bihar State Centre, Patna, India
sk.shrivastava@nic.in
[2] Department of Computer Science, Government Arts College,
Udhagamandalam, The Nilgris 643002, India
sk.mahendran@yahoo.co.in
[3] Nalanda Open University, 3rd Floor, Biscomaun Bhawan,
Gandhi Maidan, Patna, India
amarnathpandey@gmail.com

Abstract. GIS based Smart Energy Infrastructure Architecture and Revenue Administration is an integrated framework of web, mobile and GIS technology to manage electrical infrastructure and produce energy bills for the consumers. This framework helps to plan new electrical transmission infrastructure needed for quality power supply and assist in detailed planning for infrastructure. This helps to create state-of-the-art Geomatics oriented models which can assist in decentralized planning and development for robust development. The Revenue Administration Model being proposed is easily adaptable which allows for ultimate flexibility as government processes may change over a time due to changes in tariff, as well as being able to easily integrate with external applications such as Revenue accounting, ATP, IVRS, GIS, Spot billing, Payment Gateway, SMS alert and work management. The framework focuses on practical steps to be carried out for exploiting full potential of technology convergence, with emphasis on technically viable smart energy infrastructure keeping in view of sustainable growth.

Keywords: GIS · Electricity · Mobile · Framework · Geographical Information Systems · SOA · Geomatics

1 Introduction

A comprehensive framework needs to be developed for smart electrical infrastructure and revenue administration for all the electrical resources. GIS mapping of various functional equipments such as Grid Sub Stations, Power Sub Stations, Power Transmission Line, and Transmission Towers need to be completed to complete the network diagram. GIS based application model is being proposed for Grid Sub Stations 220 kV/133 kV/33 kV/11 kV, Transmission Line (Operational, Proposed, Under Construction), Transmission Towers along with attributes, Power Sub Stations etc. The application model must provide facility for buffer analysis of these substations based on

© Springer Nature Singapore Pte Ltd. 2016
S. Subramanian et al. (Eds.): CSI 2016, CCIS 679, pp. 237–246, 2016.
DOI: 10.1007/978-981-10-3274-5_19

Fig. 1. Grid substations with their attributes

capacity for 50 km, 25 km, 5 km etc. which may to identify those habitations which are not yet served. Distribution of electricity facilities can be mapped on GIS platform as shown in Fig. 1. This helps to spatially analyze new facilities to be created for extending quality electricity to the citizens [1].

For effective energy management an comprehensive energy billing software need to be developed for accepting various modes of payments, delivering services through multi-channel delivery systems and allowing customers access to their current billing and accounts information. It is also necessary to enhance security as the accessibility may be made open through mobile phones and there is need for configuring high end security and access rules for individual consumers. The services can be accessible either through mobile APP or through SMS and all the stakeholders need to access the updated information instantaneously. There is are need to develop a mobile module which is easily configurable, and allows maximum flexibility as government processes may change due changes in government rules over a time. There is also need to integrate itself with external platforms or applications such as Financial Accounting, Any Time Money, Any Time Payment, Interactive voice Response System, GIS and work and asset management systems. These is need to create a comprehensive complaint management system to address various grievance at different levels. The integrated electricity ICT based framework proposed should help in building, maintaining and operating the functions in an efficient and coordinated manner. The methodology is economical as well comply with inter-State transmission system or intra-State transmission also complies with the directions of the Regional Load Dispatch Centre and the State Load dispatch Centre. These are existing problems associated with planning in distribution system which can be solved using new ICT tools [7, 14] based on GIS and Mobile technology because there is necessity of accurate and up-to-date information of the assets of electricity network. GIS technology can help to discover new things about infrastructure planning and various kinds of financial investments as well as various project risks associated and also allows to simultaneous assess technical, financial, and environmental factors for enhancing the networks. It is necessary to align with existing habitations to ensure that electricity networks do not affect the populace.

2 GIS Modeling of Electricity Networks

Over Years GIS has been used in electrical domain to develop various kinds of systems to fetch huge database information and understand behavior of customers and also understand various other aspects of electrical management such as billing, material management, inventory management, analysis of distribution networks and outage reporting. Geographical Information Systems are now being used currently for the mapping and modeling of distribution network systems. GIS software [5, 15] can help to model changes in the electricity network, install new equipments, cover m ore area, enhance services and various parameters in the network can be updated in less time through mobile phones so that accurate information on regular interval can be collected and integrated into a comprehensive system. GIS mapping process starts with mapping of infrastructure and a visual interface to the spatial data as well as attribute data can be created. The organization of complete electrical infrastructure is depicted in Fig. 2.

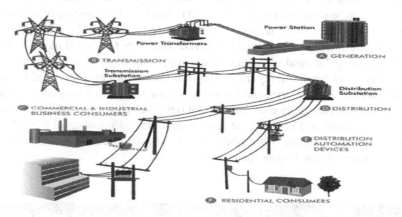

Fig. 2. Power transmission network

This kind of web GIS application not only support normal database queries, spatial queries but information can also be examined through a variety of spatial attributes such as distance, proximity, and elevation. Geographical Information System can also help in network route planning which may determine the path based on optimality principle and provides the shortest distance and at minimum cost. Spatial data analysis of electrical infrastructure can help stakeholders to analyze various kinds of patterns in spatial data and MIS. Business Intelligence can also be developed as decision support system with the purpose of improving network efficiency [2, 11] of electrical services, delivery services to the citizen and customer grievances management. Mapping and digitization of electrical transmission network, Consumer geo-tagging and imposition of the various geospatial base map of the identified location such as habitation, road, industries, schools and other infrastructures with related important attribute data on available satellite imagery [4] has been extensively being used as excellent practice of delivering quality services in many of the service applications. However, the main challenge is to integrate GIS components with main distribution processes like Management of new Connections, establishment of additional networks, electricity networks for rural areas, Collection of Payments from consumers, other Customer Services, Network analysis, Network Coverage analysis, Habitation coverage Analysis, area uncovered analysis etc. This methodology emphasizes that there is need for adoption of Service-oriented architecture (SOA), Mobile Computing, SMS and using standard middleware tools with business rules for integrating Web GIS applications with other applications.

3   Geospatial Model for Power Transmission Line

Geomatics-based Application Model for Electricity network can be conceived as an enterprise G2G/G2C based decision-support system which uses Distributed Architecture and is based on Service Oriented Architecture. Spatial data related to electricity network need to be collected about distribution network [3, 8] consisting of information related to power substations, feeders, revenue village locations, habitations as well as other basic amenities associated with the citizen. The entire GIS based framework may assist in planning for rural electricity related e-governance services up-to panchayat level and providing connectivity through optical fiber cables over electrical transmission networks. Electricity transmission poles are integral part of integrated electrical infrastructure and each poles contains different type of electrical equipments. Inventory of these equipments and maintenance require up to date information of these devices along with location. Figure 3 describe inventory at each electrical transmission pole which is suitable for up to date status of equipments.

Fig. 3. GIS mapping of electricity transmission poles

The digitization of electrical related assets of network management, consumer indexing and mapping involves the following steps:

- GPS survey of electrical consumers and network assets.
- Digitization of electrical network assets (Substations, Feeders, Transformers and Poles)
- GIS mapping, indexing and codification of electrical consumers and network assets with defined electrical relationships:

The applications of GIS are important for power system planning, analysis and control. They can help to improve power system visualization by depicting spatial data with transmission network and other associated assets of the electrical network. This also helps to maintain inventory of electrical items at each pole and power sub-stations.

4 Proposed Methodology for Infrastructure Mapping

GIS mapping of power distribution network requires GPS survey of infrastructure and geo referencing and mapping the relevant electrical assets on the digital base map. This can be done through a mobile App which can be used by field officers to visit each of the locations of GSS, PSS, electrical poles etc. These data then can be uploaded on to the GeoDB through Web services. A mobile application to capture current status of equipments at each electrical transmission can be developed as shown in Fig. 4.

Fig. 4. Mobile application for capturing attribute data

The data so collected can then be plotted on to GIS platform [7, 13]. There can be different kinds of deviations which can be corrected at later stage and attributes collected through mobile phone can be integrated. In some of the GIS applications various kinds of electricity consumers are also mapped to the corresponding electricity network. The purpose of these kinds of applications is to index all the consumers based on various criteria and later categorize them to create the complete geo-tagged consumer database with respect to their unique electrical address. A successful GIS framework can seamlessly [10] integrate with the spatial data with various distribution and consumer related applications such as Customer Information System, Assets Management, Outage Management [6, 9] and Utility Billing System. This can also provide interfaces for the cross-application and support data portability. The total integration of these electricity networks in only one shared physical electricity infrastructure, with copper cables for electric power and fiber optics for telecom, would result in an unprecedented improvement in e-governance infrastructure [5]. This can also help in much needed improvement in digital service delivery to citizens and bandwidth requirements for connecting remotest villages. There is growing need for bandwidth for both G2G and G2C services and expected that the entire governance services is likely to go digital.

5   Revenue Administration

A comprehensive billing software can be developed for billing the consumer based on prevailing tariff however there is growing requirement for providing spot billing facility so that consumer need to visit the electricity counters for receiving Bills as well as making Payments. Spot Billing Mobile App [3, 6] can have two separate applications which can be given to Meter reader. A Web site may be maintained for providing administrative functions. From the beginning of the day, a android mobile phone with a route chart which contains route of consumer houses which has to be covered within a day can be provided. Figure 5 depicts various facilities to be provided to the meter readers for generating energy bills, taking geo-tagged, time-stamped photograph of meters and synchronizing with servers.

Fig. 5. Mobile application for revenue billing

The Meter Reader can preload the entire details of the electricity bill details in the mobile before moving to the location. If the particular meter in the route map is indicated with the green color indicate that data is to be collected and in case the color is red then it indicates that data has already been captured in earlier visit. This helps in capturing all households and no household will be left. Meter readers need not do any calculation manually rather it can be done through the mobile app. All that Meter

Reader need to do is to get the meter reading and send it to the Server using mobile app as web service. There is also possibility that SMS service can be enabled for this. Then the billing software running on the server does the entire calculation and processed electricity bills are may be sent to the relevant consumers via SMS or Consumers can also view the electricity bills using Mobile App. It is also possible to lodge complain by the meter reader and report fault in case it exists.

6 BharatNet over Electricity Poles for Remote Connectivity

A GIS based electricity network map of most of the electricity infrastructure including Grid Sub Station (GSS), Power Sub Station (PSS), HT Line, LT Line, pole, transformer etc. may be created. The locations and attributes of these infrastructure may be captured through mobile App. This can use various GPS points taken of Grid Sub stations (GSS) and Power sub stations (PSS) and even towers of the electrical transmission lines along with various attributes. Various layers can be prepared for mapping every aspect of transmission lines [2] and mapping the path by which they can reach up-to the Block HQ. From Block HQ to Panchayat electrical infrastructure of 33 kV or 11 kV can be used. This GIS mapping is part of Digital India initiative [15] in which optical fiber cable is to be taken to panchayat over electrical transmission lines.

GIS Database Development:

- Digitization Process:
  - GPS survey of electrical consumers and network assets.
  - Digitization of electrical network assets.
  - GIS mapping, indexing of electrical consumers and network assets
- Data Collection:
  - GPS Base Station & adequate number of GPS Receivers.
  - Surveyors walk along and capture the spatial position.
  - Collect attribute data.
  - The digital base map must show the important landmarks for better visualization.

The process of integration of electricity transmission network initially excel sheet need to be prepared containing all the required locations and their associated parameters. An android application may also be developed for capturing geo-tagged location photographs and other attributes. The master data available can be downloaded to the mobile phone so that less work is to be done by the field engineer and also quality of Data can improve [12]. The longitude and latitude which is one of the important feature can be converted from degrees, minutes seconds to decimal degrees and then spatial data can be imported in GIS software for creating electricity network. Later the data can be ported to Web GIS platform so that it is possible to view the network over internet or intranet. This can also lead to creation of coordinated network.

7 Technology Perspective

GIS enhance visualization of power systems by associating spatial data with transmission assets, such as animation, making them attractive platforms for displaying geographically referenced real time power system data such as the voltage and line loading monitoring [2]. GIS information is stored in geographical map layers making it easy to relate transmission network conditions with other relevant information such as weather, vegetation growth, and road networks. Real-time weather data integrated in GIS increases the operator's situational awareness. For example, with the help of such a system, the identification of a weather front moving towards a given area enables operators to quickly determine transmission facilities with increased risks of outage. Figure 6 represents integration framework of GIS technologies to support decision making process.

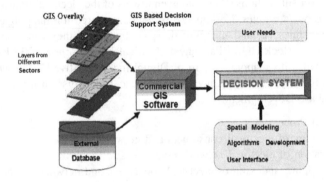

Fig. 6. GIS integration framework and spatial modeling

8 Optimizing Electrical Lines Routing and Load Forecasting

The electrical transmission network routing and optimization is highly complex procedure involving lot of parameters to be considered before laying the design. In case it is not planned properly it may lead to health issues of citizen due to the electric and magnetic fields, especially from high voltage transmission lines. The cost involved in shifting transmission line from one location to another aligning to habitation is a difficult task. During the selection of route for electricity transmission line, a straight route with minimum level of curve [4, 13] is desirable as it gives the best engineering and economic solution. However in order to connect to various habitations the electricity network need to be created near habitation only. In order to achieve this electrical route the line may have to pass through certain places which are already inhabited by citizen or areas that are unsuitable for locating the transmission towers. GIS technology can be used to analyze the selection of suitable areas for transmission lines [7, 10], so that there is minimal disruption and less hazard to the citizen. This can be done by methods such as minimizing the number of trees in a forest area, implement optimal routing algorithms based on electrical and material properties in addition to location characteristics, visualize the network on a map helps make appropriate decision. Installing transmission lines is very expensive, so it's not an option to make errors about location. Buffer zone concept from spatial informatics can help in routing the High tension transmission line near to a populated area, where spatial buffer zone will protect the inhabitants from strong electric and magnetic field effects. Forecasting the amount of future load growth and predicts the location of load increment is called load forecasting. It is very important for power planning and the whole planning work depends on it. To perform spatial load forecasting techniques Gathering spatial information can be performed on the Geographic Information System (GIS) Platform.

9 Conclusion and Future Work

Applications of GIS in Network Management System as stated in this paper, which are; Integrating between text and spatial data can be used to perform auditing of Energy, Management of Load, Planning for Network expansion, Network analysis, determining the optimum, shortest, and most economic path for electrical transmission lines; forecasting and predicting the amount of power needed in the future [6] etc. Geographical Information System in electrical power management system can help to determine the optimal path for creating or expanding transmission lines. This can also help to forecast the growth of load and increase in infrastructure requirements due to increase in services and manage electrical infrastructure in case of disasters. The framework can be extended in future to focuses for exploiting full potential of technology convergence, with emphasis on technically viable smart energy infrastructure keeping in view of sustainable growth.

Acknowledgement. This work is a part of the National GIS Framework being established by NIC for 1:10 K scale multi-layer GIS project and iBhugoal-Phase-II project funded by Govt. of Bihar.

References

1. http://egov.eletsonline.com/2014/02/improving-transparency-and-delivering-quality-services-with-ibhugoal/
2. http://indianpowersector.com/home/2013/05/it-interventions-in-power-distribution-reforms-in-india-adoption-of-new-technologies-and-integration-challenges/
3. http://nisg.org/files/documents/UP1418304168.pdf
4. iBhugoal Project Case Study by DARPG, Government of India. http://www.darpg.gov.in/sites/default/files/iBhugoal_case.pdf
5. iBhugoal. http://gis.bih.nic.in/
6. Sinha, J.: GIS Application in Power Distribution Utility, UPCL, Dehradun
7. Murata, M.: A GIS application for power transmission line. In: ESRI User Conference Proceedings, US (1995)
8. Rezaee, N., Nayeripour, M., Roosta, A., Niknam, T.: Role of GIS in distribution power systems. World Acad. Sci. Eng. Technol. **3**, 36 (2009)

9. Nagaraja Sekhar, A., Rajan, K.S., Jain, A.: Application of GIS and spatial informatics to electric power systems, IIT Bombay (2008)
10. National GIS Framework Document of National Informatics Centre, Government of India
11. Saheed Salawudeen, O., Rashidat, U.: Electricity Distribution Engineering and GIS (2006)
12. Kong, P.: The design and implementation of JiNing electric power company's electric transaction MIS. ShanDong University (2007)
13. Smith, P.H.: Electrical distribution modeling, MS thesis, Blacksburg, Virginia (2005)
14. Shrivastava, S.K., Pandey, A.N., Kumar, P.: Geomatics oriented model for converging electrical transmission networks into state wide area networks. In: 2014 International Conference on Computing for Sustainable Global Development (INDIACom), pp. 75–80 (2014). doi:10.1109/IndiaCom.2014.6828055
15. www.engineering.nottingham.ac.uk/icccbe/proceedings/pdf/pf63.pdf

An RFID Cloud Authentication Protocol for Object Tracking System in Supply Chain Management

S. Anandhi[✉], R. Anitha, and Venkatasamy Sureshkumar

Department of Applied Mathematics and Computational Sciences,
PSG College of Technology, Coimbatore 641004, India
san@amc.psgtech.ac.in

Abstract. Radio Frequency Identification (RFID) is a valuable technology for tracking objects in the supply chain. Security and privacy requirements arise with the fast deployment of RFID in supply chain in a heterogeneous environment. Authentication is one of the important security requirements in cloud environment. Even though several RFID cloud authentication protocols are available for supply chain management, they lack to satisfy some security requirements. There is a need for secure, efficient, and scalable protocol for agile supply chain. In this paper, an RFID cloud authentication protocol is proposed and an informal security analysis is carried out. Performance analysis is done with respect to the tag entity. The proposed protocol is scalable and it preserves tag/reader privacy, provides mutual authentication and resistant to many attacks. Comparison with the existing protocol in terms of communication cost shows that our protocol outperforms the other protocols.

Keywords: Mutual authentication · RFID · Cloud authentication · Supply Chain Management · Security and privacy issues

1 Introduction

In Supply Chain Management (SCM) system, products are moved from supplier to customer. It integrates and coordinates the flow of material and information from supplier to manufacturer to wholesaler to retailer and to consumer. Effective SCM ensures the availability of the product when needed. SCM is a complex and knowledge intensive process [3,15]. In order to address the inventory problems and logistics problems in SCM, the products are to be tracked to check its availability in an efficient manner. Radio Frequency IDentification (RFID) technology implemented in supply chain ensures that the right goods are available in the right place. RFID makes the supply chain considerably more precise and improves the efficiency and reliability of the entire chain.

An RFID system has three components tag, reader, and server. The tags are classified as active tag and passive tag. Active tag has read/write capabilities, where as passive tag does not contain internal power source and the tag can be

© Springer Nature Singapore Pte Ltd. 2016
S. Subramanian et al. (Eds.): CSI 2016, CCIS 679, pp. 247–256, 2016.
DOI: 10.1007/978-981-10-3274-5_20

read only at very short distances [12,13]. An active tag consists of the product details and secret data for secure communication. Readers provide linkage between the tag and the server, allowing data to be read from the tag and transmitted to the server. Reader scans the tag to check the data and communicates with networked servers and reader can be a mobile or handheld device. The server is responsible for storing and processing the data. The tag data can be updated by a reader [8,15].

After a product is manufactured, a unique RFID tag is attached. The product moves to warehouse, retail store and finally to customer. In warehouse, the object availability is checked with RFID tag. Customer uses RFID tag attached to the product in any novel RFID application. After sales also the service provider records the history of the product service details in the tag. Therefore, it is necessary to track the object till its lifetime.

Cloud environment makes supply chain to track the product throughout its lifecycle easily. As real time information is made available, administration and planning processes can be significantly improved. Cloud computing provides data availability and supports business agility [3–5]. In any RFID based system in the cloud, authentication is the most important security requirement. Each participating entity in the system can make sure that they are communicating to the right entity. Improper authentication can result many attacks like, tag impersonation, reader impersonation, and replay attack. Hence it is necessary to have protocol for achieving mutual authentication between the reader and the tag [14,20]. Thus, security and privacy issues remain a major issue [9,10]. The communication between the tag and the reader is in wireless environment and reader to cloud server is in public channel. Authentication between the reader and cloud server is achieved using Message Authentication Code (MAC) [6,7].

In this paper, an enhanced RFID cloud authentication protocol is proposed to provide solutions for security and privacy issues. Cloud server stores the tag details and its associated reader details in a secured manner, which enables the application logic to change the reader and the tag details in a pervasive environment. The main objective of this paper is to highlight the security and privacy issues and propose an enhanced RFID authentication protocol for the supply chain management with greater scalability.

The rest of the paper is organized as follows. Section 2 details some related work. Proposed protocol is presented in Sect. 3. Section 4 describes the security and performance analysis of the protocol. In Sect. 5, the paper is concluded with some future work.

2   Related Work

Initially, RFID security protocols are modelled to address mutual authentication as an important security requirement [11,16]. As mobile readers are introduced, new requirements for efficient protocol are raised and also the design should be made privacy preserving. Existing protocols are not well suitable for addressing several security and privacy issues [10]. It is very difficult for companies to maintain the storage and authentication services. Evolution of cloud, serves the purpose. Implementation of RFID for supply chain management system in cloud environment brings out new security challenges in the research field like ownership transfer, authority recovery, data usage, etc.

In [3], the authors aimed to provide mutual authentication between the tag and the reader in SCM. They have maintained routing table in the tag which is prone to attack. Once the authentication is successful, the tag is ready to receive the communication from the other reader in the chain. It is very difficult to maintain the secret value in all the readers in a heterogeneous environment. Authority recovery is not addressed.

In [20], encrypted hash table is introduced to preserve the privacy of the tag/reader from database keepers in cloud. Scalable protocol is proposed. The main important assumption in that paper is communication between reader and cloud provider is through Virtual Private Network (VPN). Cost of deploying VPN is not suitable for small and medium enterprises and business data is exposed to VPN agency. In this protocol, tag initiates the protocol execution where as in all other protocols reader initiates. It is advisable that always reader initiates the execution. Otherwise, the tag possibly will communicate with an unauthorized reader. Any reader deployed in the supply chain can perform authentication. Updation of reader information in tag is not addressed.

In [10], mobile reader is considered in the design. This protocol makes the tag free from storage and computations, but heavy computation on reader device. As a mobile reader it is difficult to have such a storage space and computational capacity. The authors ensure the data integrity between the tag and reader for every message using hashing. Between the reader and cloud server MAC is used. Reader can say the missing products as available one, which will create serious problem in inventory tracking. The authentication protocol fails accountability property since, without the tag's knowledge the reader can be able to prepare the message [17,18]. The formal security verification is carried out using Automated Validation of Internet Security Protocols and Applications (AVISPA) tool [1,2,19]. The various technical issues observed in this protocol are as follows:

- The mobile reader initiates the protocol, the tag generates a random number and prepares M_1 and sends without authenticating the reader. If replay attack is carried out, then the tag repeats many computations unnecessarily.
- In case of tag impersonation attack, fake tags can store details in the reader which leads to memory overflow, a kind of DoS attack.
- The cloud server encrypts the session key and tag ID and sends it to the reader, but that could be prepared by the reader itself and freshness property is not proved.

In [14], security requirements, common attacks and privacy problems in RFID authentication protocol are defined. The message transferred in this protocol is protected with cryptographic primitives. Trusted Third Party (TTP) is included for ownership transfer, authority recovery, and data sharing. In the tag, secret key is stored and updated for every successful session but it is not used for encryption. This protocol cannot withstand DoS attack; however privacy is still preserved.

Almost all the authentication protocols meet the basic authentication purposes. But, when value added services in supply chain domain are considered, the authentication protocol has to be redesigned by defining the security and privacy requirements clearly. It is essential to design an efficient RFID authentication protocol to resist all possible attacks and threats and implementation ease.

3   Proposed Authentication Protocol

In the supply chain management the objects are embedded with passive tag. The reader devices in various departments are used to track the objects. This front-end communication uses public radio channels. As the reader is not capable of storing large amount of data, storage is moved to the cloud server. This back-end communication is to be secured as the database service provider is not a trusted one. The objective of cloud deployment is to provide scalability and the ability to manipulate the tag data in a secure manner by different supply chain participant. When the object is moved from one department to another, the reader associated with the tag is changed. The application logic written in the server updates the active reader, which is currently associated with the tag.

The proposed authentication protocol is depicted in Fig. 1. Cloud server uses Encrypted Hash Table (EHT) to store tag ID and its associated details like the owner of the tag and list of readers [20]. The protocol preserves the list of authenticated readers for the tag. It is made easy to control the legitimate readers processing on the tag and delegating the reader responsibility to the other readers in the list. The notations used in this protocol are shown in Table 1.

3.1   RFID Cloud Authentication Protocol

The proposed protocol consists of four important online phases.

(1) Setup phase. The tag is associated with the reader and the details are stored in the cloud server. Tag possesses TID, pre-computed $h(OW \parallel TID)$, reader identity R, and pseudo identity T_{id}. The reader holds its permanent identity R. In cloud server EHT is stored and its structure is as follows. For each tag separate record is maintained. Its index value is $h(OW \parallel TID)$. Each record comprises list of readers associated with the tag in the past, and active reader detail $E(T_{id} \parallel R)$. Among the list of readers only one can be active at a time which is decided by the application logic. The list is encrypted by the reader to achieve privacy of the reader tag association.

(2) Reader authentication phase. Reader is authenticated to the tag in this phase to ensure that the tag responses only to the authenticated reader.

Step RAP1. The reader (R) selects a random nonce r_1 and initiates the communication by sending $m_1 = \langle r_1, h(r_1 \parallel R) \rangle$ to the tag.

Table 1. Notations used in the protocol

Notation	Description
Tid	Pseudo id, updated upon successful authentication
TID	Manufacture's tag ID
R	ID of the reader
r_1, r_2	Random nonce generated by reader and tag
K_{RC}	Key between reader and cloud server for MAC
$H_{K_{RC}(m)}$	Message Authentication Code of m prepared using K_{RC}
TS	Timestamp
T	Acceptable time difference between the tag's response and cloud server's response at reader
$\parallel$	Denotes concatenation of two strings
EHT	Encrypted Hash Table contains the hashed index and list of encrypted reader tag details
$h(\cdot)$	Secure one way hash function
$E(\cdot)$	Encrypt function by using the reader's secret key
$D(\cdot)$	Decrypt function by using the reader's secret key
$\oplus$	Denotes bitwise exclusive OR operation
T_1	Timestamp at tag side
T_2	Timestamp at cloud server side
OW	Owner of the tag

Step RAP2. After receiving m_1, tag computes $V_1 = h(r_1 \parallel R)$ and checks $V_1 \overset{?}{=} h(r_1 \parallel R)$ to verify R. If it fails, then authentication is not done and hence, the tag terminates the session.

Step RAP3. Otherwise, tag generates a random nonce r_2 and computes $V_2 = h(T_{id} \parallel R \parallel r_1)$. Now, the tag sends a message m_2 which contains $h(OW \parallel TID)$ (pre computed and stored in the tag), V_2, r_2 and the current timestamp T_1 to the reader.

(3) Tag authentication phase. In order to authenticate the tag, reader communicates with cloud server.

Step TAP1. Upon the receipt of m_2, the reader stores $\langle V_2, r_2, T_1 \rangle$ corresponding to r_1. The reader computes $L_1 = H_{K_{RC}}(h(OW \parallel TID))$, which is a MAC for the message $h(OW \parallel TID)$ and sends the message $m_3 = \langle h(OW \parallel TID), L_1 \rangle$ to the cloud server.

Step TAP2. After receiving the message m_3, the cloud server computes $L_1^* = H_{K_{RC}}(h(OW \parallel TID))$ and checks $L_1^* \overset{?}{=} L_1$. If it is not true, then authentication fails. Otherwise, cloud server checks EHT table entry for the index value $h(OW \parallel TID)$ and finds the active reader R with its associated

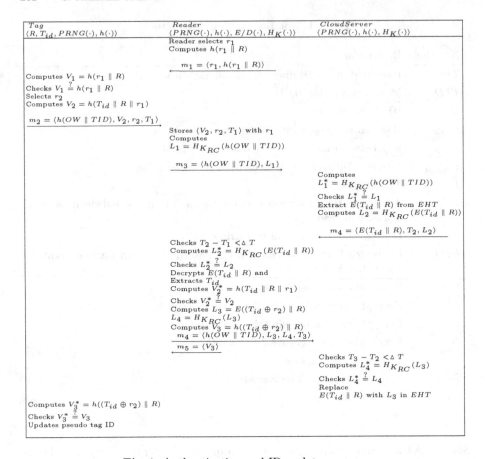

Fig. 1. Authentication and ID update process

$E(T_{id} \parallel R)$. If not found, then authentication fails. Also, the cloud server computes, $L_2 = H_{K_{RC}}(E(T_{id} \parallel R))$ and sends $m_4 = \langle E(T_{id} \parallel R), T_2, L_2 \rangle$ to the active reader.

Step TAP3. Reader R checks the time delay $T_2 - T_1 <_\Delta T$, where ΔT is the acceptable time delay. If the time difference is too long, then it is identified as replay attack and authentication fails. Otherwise, R computes $L_2^* = H_{K_{RC}}(E(T_{id} \parallel R))$ and also checks the correctness of $L_2^* \overset{?}{=} L_2$. If it is not true, then authentication fails.

Step TAP4. Now, the reader decrypts the cipher $E(T_{id} \parallel R)$ for the extraction of T_{id}. The reader computes $V_2^* = h(T_{id} \parallel R \parallel r_1)$ and checks computed V_2^* with the stored V_2. If the verification $V_2^* \overset{?}{=} V_2$ is successful, then the reader confirms that the tag is an authenticated one.

(4) Pseudo Tag ID updation phase. To preserve the privacy of the tag, its identity is to be updated upon every successful authentication. Anonymity is also achieved by using dynamic ID.

Step TUP1. The reader computes $L_3 = E((T_{id} \oplus r_2) \parallel R)$, $L_4 = H_{K_{RC}}(L_3)$ and sends $m_4 = \langle h(OW \parallel TID), L_3, L_4, T_3 \rangle$ to the cloud server.

Step TUP2. Before updating the record in the EHT table, cloud server checks $T_3 - T_2 < \triangle T$ to prevent replay attack. The cloud server computes $L_4^* = H_{K_{RC}}(L_3)$ and checks $L_4^* \overset{?}{=} L_4$, if not terminates the updation process. Otherwise, cloud server replace the existing $E(T_{id} \parallel R)$ with L_3 in EHT table corresponding to the index $h(OW \parallel TID)$. It is worth noting that, the old T_{id} in $E(T_{id} \parallel R)$ is replaced by $T_{id} \oplus r_2$ in the updation process.

Step TUP3. At the same time, the reader also computes $V_3 = h((T_{id} \oplus r_2) \parallel R)$ and sends the update $m_5 = \langle V_3 \rangle$ to the tag. The tag computes $V_3^* = h((T_{id} \oplus r_2) \parallel R)$ and checks whether the received message $V_3^* \overset{?}{=} V_3$ and updates the pseudo tag ID.

Upon successful authentication, the access control policies can be executed.

4   Analysis of the Protocol

In this section, security analysis, performance analysis and applicability issues of the proposed protocol are discussed.

4.1   Security Analysis

In this section, the protocol is analysed for various security issues informally.

Mutual authentication. Tag checks $h(r_1 \parallel R)$ from the request message and authenticates R. Reader decrypts $E(T_{id} \parallel R)$ which is received from the cloud server to check whether the tag is associated with it or not. By computing $h(T_{id} \parallel R \parallel r_1)$ it checks with message V_2 in **Step TAP1** and authenticates the tag. The reader and the cloud server authenticate each other by using Message Authentication Code (MAC). Thus mutual authentication between the entities is obtained.

Replay attack. Attacker performs replay attack in three ways: (i) The replay of reader's message in **Step RAP1** will not be authenticated by tag because the message contains random number r_1, which will be newly generated by the reader for each communication. (ii) The replay of tag's message in **Step RAP3** will not be possible because updated T_{id} is used for each session and also a random number r_2 is a part of the message. (iii) The replay of cloud server's message in **Step TAP2** is not possible because of timestamp T_2 inclusion.

Untraceability. The data stored in the cloud server is encrypted and it is difficult to map which tag is associated to which reader. The tag ID and reader

ID are also securely transmitted. Pseudo ID for the tag T_{id} is updated upon successful authentication. Thus, tag cannot be traced by attackers.

Tag impersonation attack. This attack can be performed by the attacker to project that an unavailable object as available or sometimes to perform denial of service. In the proposed protocol, tag authenticates the reader only then it sends response to the reader. **Step RAP2** fails for fake tag. Thus tag impersonation is not possible.

Reader impersonation attack. Request from the forged reader is not processed by the tag because it is not possible to generate $h(r_1 \parallel R)$ by a forged reader.

Forward and Backward secrecy. From a compromised tag, the attacker obtains the data stored in it. But it is not possible to trace the previous and future conversation with the current data. In the proposed protocol, T_{id} is updated for each successful authentication. It is difficult to predict the previous and future T_{id}, because it is updated by Xoring a random nonce r_2 with T_{id}.

4.2   Performance Analysis

In RFID cloud authentication protocol, tag is a lightweight component, so the protocol designed is suitable and efficient with respect to tag entity. We consider the operations performed by the tag and secret details stored in the tag for discussion. The performance factors and their values are listed in Table 2. It is observed that the number of hash operations to be performed by the tag is less still the protocol meets the security needs. Message prepared by the tag in the protocol execution is one. Privacy is also preserved by not revealing the tag or reader identity.

Table 2. Comparative analysis on performance properties

Performance measure	Bi and Mu [3] (2010)	Xie et al. [20] (2013)	Lin et al. [14] (2015)	Dong et al. [10] (2015)	Cao et al. [4] (2016)	Proposed protocol
M_1	1	1	2	1	1	1
M_2	3	2	4	3	4	3
M_3	2	2	1	1	2	1
M_4	4	3	4	2	3	3
M_5	5L	9L	4L	8L	9L	6L
M_6	$h(\cdot), PRNG(\cdot)$	$h(\cdot), PRNG(\cdot)$	$h(\cdot), \oplus, PRNG(\cdot)$	$h(\cdot), PRNG(\cdot)$	$h(\cdot), PRNG(\cdot)$	$h(\cdot), \oplus, PRNG(\cdot)$

M_1: Random number generation in Tag, M_2: Number of hash operations, M_3: Number of message sent by the tag, M_4: Number of secrets on tag, M_5: Length of the hash values transmitted, M_6: Cryptographic operations by Tag.

In order to ensure the freshness of the message sent by the tag, at least one random nonce is generated for each session. Except Lin et al.'s [14] protocol, all other protocols generate one random nonce. Number of hash operations and number of messages sent by the tag is minimized, except for the Xie et al.'s [20] protocol, because their assumption is that the communication between the reader and cloud server is through VPN. Considering L as the length of hash values transmitted in the protocol, proposed protocol uses $6L$ for successful authentication, whereas the recently proposed protocol by Cao et al. [4] uses $9L$. Thus, the proposed protocol provides security with less communication cost.

4.3 Applicability and Complexity

The proposed protocol requires tag to support PRNG for maintaining the freshness of the message sent from the tag. It requires lightweight one way hash function to achieve data integrity. In order to secure communication between the reader and cloud server, readers are required to support symmetric encryption and decryption such as AES algorithm. Reader executes $PRNG(\cdot)$, $h(\cdot)$, $XOR(\cdot)$, and a MAC algorithm. Complexity to identify the tag by the reader lies only on the search logic in the cloud server. Thus search is scalable. The random numbers involved in the message, which is hashed provides freshness for each message transmitted in the protocol. Privacy of the tags and readers are preserved against attackers as well as from the cloud service providers.

5 Conclusion

In this paper, we have briefly analysed many recent RFID cloud authentication protocols for SCM. Cross-organizational information sharing, change in business policies and massive scale of RFID related information flow creates security and privacy challenges in a rapid manner. Entities participating in the protocol execution are from heterogeneous environment, which makes the protocol execution more challenging. This paper provides an enhanced authentication protocol which meets the security needs. The future work includes design of cloud based ownership transfer protocol with formal security analysis. In all authentication protocols the reader, tag, and its association must be known in advance. But, in case of Internet of Things (IoT) environment, it is very challenging to make the initial setup.

References

1. Amin, R., Islam, S.H., Biswas, G., Khan, M.K.: An efficient remote mutual authentication scheme using smart mobile phone over insecure networks. In: International Conference on Cyber Situational Awareness, Data Analytics and Assessment (CyberSA), pp. 1–7. IEEE (2015)
2. Amin, R., Islam, S.H., Biswas, G., Khan, M.K., Kumar, N.: An efficient and practical smart card based anonymity preserving user authentication scheme for tmis using elliptic curve cryptography. J. Med. Syst. **39**(11), 1–18 (2015)
3. Bi, F., Mu, Y.: Efficient RFID authentication scheme for supply chain applications. In: 2010 IEEE/IFIP 8th International Conference on Embedded and Ubiquitous Computing (EUC), pp. 583–588. IEEE (2010)

4. Cao, T., Chen, X., Doss, R., Zhai, J., Wise, L.J., Zhao, Q.: RFID ownership transfer protocol based on cloud. Comput. Netw. **105**, 47–59 (2016)
5. Cao, T., Shen, P., Bertino, E.: Cryptanalysis of some RFID authentication protocols. J. Commun. **3**(7), 20–27 (2008)
6. Chaudhry, S.A.: A secure biometric based multi-server authentication scheme for social multimedia networks. Multimedia Tools Appl. **75**(20), 12705–12725 (2015)
7. Chaudhry, S.A., Mahmood, K., Naqvi, H., Sher, M.: A secure authentication scheme for session initiation protocol based on elliptic curve cryptography. In: IEEE International Conference on Computer and Information Technology; Ubiquitous Computing and Communications; Dependable, Autonomic and Secure Computing; Pervasive Intelligence and Computing, pp. 1960–1965. IEEE (2015)
8. Chen, S.M., Wu, M.E., Sun, H.M., Wang, K.H.: CRFID: An RFID system with a cloud database as a back-end server. Future Gener. Comput. Syst. **30**, 155–161 (2014)
9. Chien, H.Y.: SASI: a new ultralightweight RFID authentication protocol providing strong authentication and strong integrity. IEEE Trans. Depend. Secur. Comput. **4**(4), 337–340 (2007)
10. Dong, Q., Tong, J., Chen, Y.: Cloud-based RFID mutual authentication protocol without leaking location privacy to the cloud. Int. J. Distrib. Sens. Netw. **2015**, 209 (2015)
11. Farash, M.S.: Cryptanalysis and improvement of an efficient mutual authentication RFID scheme based on elliptic curve cryptography. J. Supercomput. **70**(2), 987–1001 (2014)
12. He, D., Kumar, N., Chilamkurti, N., Lee, J.H.: Lightweight ECC based RFID authentication integrated with an ID verifier transfer protocol. J. Med. Syst. **38**(10), 1–6 (2014)
13. He, D., Zeadally, S.: An analysis of RFID authentication schemes for internet of things in healthcare environment using elliptic curve cryptography. IEEE Internet Things J. **2**(1), 72–83 (2015)
14. Lin, I.C., Hsu, H.H., Cheng, C.Y.: A cloud-based authentication protocol for RFID supply chain systems. J. Netw. Syst. Manag. **23**(4), 978–997 (2015)
15. Reyes, P.M.: RFID in the Supply Chain. McGraw-Hill Professional, Pennsylvania (2011)
16. Sundaresan, S., Doss, R., Zhou, W.: Zero knowledge grouping proof protocol for RFID EPC C1G2 tags. IEEE Trans. Comput. **64**(10), 2994–3008 (2015)
17. Sureshkumar, V., Anitha, R., Rajamanickam, N., Amin, R.: A lightweight two-gateway based payment protocol ensuring accountability and unlinkable anonymity with dynamic identity. Comput. Electr. Eng. (2016, in press). http://dx.doi.org/10.1016/j.compeleceng.2016.07.014
18. Sureshkumar, V., Ramalingam, A., Anandhi, S.: Analysis of accountability property in payment systems using strand space model. In: Abawajy, J.H., Mukherjea, S., Thampi, S.M., Ruiz-Martínez, A. (eds.) SSCC 2015. CCIS, vol. 536, pp. 424–437. Springer, Heidelberg (2015). doi:10.1007/978-3-319-22915-7_39
19. Viganò, L.: Automated security protocol analysis with the AVISPA tool. Electr. Notes Theoret. Comput. Sci. **155**, 61–86 (2006)
20. Xie, W., Xie, L., Zhang, C., Zhang, Q., Tang, C.: Cloud-based RFID authentication. In: 2013 IEEE International Conference on RFID (RFID), pp. 168–175. IEEE (2013)

Graph Cut Based Segmentation Method for Tamil Continuous Speech

B.R. Laxmi Sree[(✉)] and M.S. Vijaya

PSGR Krishnammal College for Women,
Avinashi Road, Peelamedu, Coimbatore 641004, Tamilnadu, India
viporala@yahoo.com, msvijaya@psgrkc.com

Abstract. Automatic segmentation of continuous speech plays an important role in building promising acoustic models for a standard continuous speech recognition system. This needs a lot of segmented data which is rarely available for many languages. As there are no industry standard speech segmentation tools for Indian languages like Tamil, there arises a need to work on Tamil speech segmentation. Here, a segmentation algorithm that is based on Graph cut is proposed for automatic phonetic level segmentation of continuous speech. Using graph cut for speech segmentation allows viewing speech globally rather locally which helps in segmentation of vocabulary, speaker independent speech. The input speech is represented as a graph and the proposed algorithm is applied on it. Experiments on the speech database comprising utterances of various speakers shows the proposed method outperforms the existing methods Blind Segmentation using Non-Linear Filtering and Non-Uniform Segmentation using Discrete Wavelet Transform.

Keywords: Speech segmentation · Graph cut · Tamil speech · Phonetic-level segmentation

1 Introduction

The success of any of the research on speech recognition system lies in the correctness of segmentation of speech. The process of segmentation is done based on the requirement and nature of the problem. Building speech models requires a lot of correctly segmented speech data to ensure the accuracy of the model. The segmentation process can be done either manually or automatically. Manual segmentation of huge database is a tiresome process. Thus there is a need of automatic segmentation of continuous speech. There are some methods available for automatic segmentation of speech worked on many different languages. A few methods have been applied to Tamil speech [8–10]. The aim of this research is to propose an efficient algorithm to segment the continuous speech in Tamil.

Many approaches like knowledge-based, feature-based, sub-band based approaches have been used in speech segmentation. In [1], 13 knowledge-based acoustic parameters and Support vector machines are used for segmenting and classifying the speech components to a broad class such as vowel, stop, fricative, sonorant consonant and silence. This performs well when compared to the HMM based segmentation that used

© Springer Nature Singapore Pte Ltd. 2016
S. Subramanian et al. (Eds.): CSI 2016, CCIS 679, pp. 257–267, 2016.
DOI: 10.1007/978-981-10-3274-5_21

39 MFCC features. A blind segmentation approach is applied [2], where the short-term FFT features are extracted and non-linear filters are applied to identify the segmental points in the continuous speech. Auditory model (AM) techniques are reported better when compared to acoustic models based on FFT, cepstrum or LPC [3]. Here Multi Level segmentation is performed by repeatedly clustering the neighboring frames based on their Euclidean distance similarity measure. This segmentation works for different speakers, vocabulary or speech with background noise. It is suggested that 8-order Discrete Wavelet Transform (DWT) better suits for speech segmentation [4]. It uses cubic spline wavelet as the mother wavelet, which is applied upon the actual speech to identify the segment points. Also many other authors have focused their work of speech segmentation using Wavelet Transforms [5–7]. An algorithm based on group delay is used in [8] for segmenting continuous speech into syllabic units, which is then subjected to a HMM for classification of syllables. In [9], short-time energy and zero-crossing rate of the speech signal are used to segment the continuous speech of Indian languages. The spontaneous speech is segmented into syllabic units using sub-band-based group delay technique [10]. It makes use of two properties for smoothing the speech signal's short-term energy function; the Fourier Transform's additive property and the Cepstrum's deconvolution property for detecting the syllable boundary. In the paper [15], a tool is developed, the ALISA tool which uses a lightly supervised method. This method aligns the speech with imperfect transcript at sentence-level. A GMM-based activity detector and a grapheme-based aligner are used for speech alignment.

Accuracy of earlier researches including automatic speech recognition and other rely on the accuracy of speech segmentation. A generalized approach that segments the speech into phonetic units that is independent of the vocabulary used is proposed, which additionally aims to improve the segmentation accuracy. This also aims in bringing out an approach that can also be used in other languages as well in the future. Graph cut based segmentation have shown better results in segmenting objects of an image or motion videos [11]. The idea of representing pixels as nodes and their similarity as edges of graph paved the motivation to use graph cut for this speech segmentation problem. Here, an algorithm which uses graph cut approach to segment continuous speech is proposed. The graph cut method represents the data points as nodes of graph and their similarity as edges of the graph. It then identifies a best splitting point to split the graph. Here, we propose a statistical approach in identifying the candidate splitting points from which the best splitting point is selected to bipartite the graph whereas [11] considers l evenly spaced points as candidate splitting points and choose the best from that candidate set. The proposed work turned out with comparable segmentation results with the existing methods. The accuracy of segmentation is measured here using precision, recall and F-score measures. The result of this graph cut based segmentation is finally compared with the existing methods Blind Segmentation using Non-Linear Filtering (BSNLF) and Non-Uniform Segmentation using Discrete Wavelet Transform (NUSDWT).

2 Graph Cut Based Segmentation

Graph cut has its roots in graph theoretic techniques. Several graph theoretical techniques have been proposed for cluster analysis one of which is graph cut approach [14]. These approaches can be applied to any problem which can be converted into a graph constructed based on the data points and its neighborhood. Here the speech segmentation problem is converted to a clustering problem considering the fact that the data points belonging to a phoneme should be placed together and similar phonemes lying apart need not be connected with an edge to the nodes of this phoneme. These techniques have been already used in image segmentation where the pixels are considered as graph nodes and their relationship with neighborhood pixels are considered as edges to construct the undirected graph [11]. Similarly, here the time frames are considered as nodes and the likelihood with its locally distributed frames as the edges of the graph.

2.1 Graph Cut

A Graph $G = (V, E)$ can be partitioned into two subgraphs $G_1 = (V_1, E_1)$ and $G_2 = (V_2, E_2)$ such that $V_1 \cup V_2 = V$ and $V_1 \cap V_2 = \varnothing$, by simply removing the edges between the sub graphs G_1 and G_2. The optimality of the bipartition of the graph relies on the selection of edges that we remove. The degree of dissimilarity between the two subgraphs is calculated as

$$cut(V_1, V_2) = \sum_{u \in V_1, v \in V_2} w(u, v) \tag{1}$$

where $w(u, v)$ is the similarity between the nodes u and v (refer Eq. 5).

2.2 Optimal Cut

Identifying the optimal graph cut is necessary to turn out with optimal segmentation of speech (Fig. 1). To achieve this, a measure of disassociation, normalized cut (Ncut) [11] is used.

$$Ncut(V_1, V_2) = \frac{cut(V_1, V_2)}{assoc(V_1, V)} + \frac{cut(V_2, V_1)}{assoc(V_2, V)} \tag{2}$$

where $assoc(V_1, V) = \sum_{u \in V_1, t \in V} w(u, t)$ denotes the sum of edge weights that connects the nodes of the graph G_1 to all the nodes in the Graph G and $assoc(V_2, V) = \sum_{v \in V_2, t \in V} w(v, t)$ is the sum of edge weights that connects the nodes of graph G_2 to each node present in the Graph G.

Similarly, the normalized association between the subgraphs G_1 and G_2 of graph G is defined as:

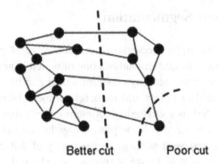

Fig. 1. Case showing the importance of optimal cut for better partition

$$Nassoc(V_1, V_2) = \frac{assoc(V_1, V_1)}{assoc(V_1, V)} + \frac{assoc(V_2, V_2)}{assoc(V_2, V)} \tag{3}$$

where $assoc(V_1, V_1)$ and $assoc(V_2, V_2)$ represents the within nodes association of V_1 and V_2 respectively. The normalized cut measure is naturally related to the normalized association, and it can be represented in terms of normalized association as follows

$$
\begin{aligned}
Ncut(V_1, V_2) &= \frac{cut(V_1, V_2)}{assoc(V_1, V)} + \frac{cut(V_2, V_1)}{assoc(V_2, V)} \\
&= \frac{assoc(V_1, V) - assoc(V_1, V_1)}{assoc(V_1, V)} + \frac{assoc(V_2, V) - assoc(V_2, V_2)}{assoc(V_2, V)} \\
&= \frac{assoc(V_1, V)}{assoc(V_1, V)} - \frac{assoc(V_1, V_1)}{assoc(V_1, V)} + \frac{assoc(V_2, V)}{assoc(V_2, V)} - \frac{assoc(V_2, V_2)}{assoc(V_2, V)} \\
&\therefore Ncut(V_1, V_2) = 2 - Nassoc(V_1, V_2)
\end{aligned}
\tag{4}
$$

Thus, performing an optimal partition of the graph can be ensured by choosing the best edge set for partition. The image segmentation problem for segmenting static images and motion sequences is considered and solved using the normalized cut criterion [11]. Here the image is represented as a graph where each pixel is represented as a node and the likelihood between the pixels as the edges. A dense stereo matching algorithm which considered the disparity between the regions in the images to segment epipolar rectified images is proposed in [12].

3   Phonetic Level Segmentation of Continuous Speech

Let the continuous speech S be denoted as feature vectors in set, $X = \{X_1, X_2, \ldots X_n\}$ where n shows the total number of frames representing the given continuous speech. A second order filter is applied to the speech wave and Discrete Wavelet Transform (DWT) is applied on the speech to extract the features of the speech. The DWT features are represented as feature vectors. A graph is build by considering each feature vector as a node and similarity between the feature vectors as the edge connecting the nodes.

The distance of the one feature vector from the other in the feature set (X) is called the physical distance. The physical distance between the nodes is considered as a factor to restrict the number of edges in the graph. An edge between nodes is allowed if their physical distance is less than the distance factor (r). The graph is then represented as a weight matrix (W), where the weight of edge is the likelihood between the feature vectors representing the nodes connected by the respective edge. The eigenvectors for the whole system represented as a graph is calculated and the eigenvectors with the median eigenvalues are chosen. Then the optimal cut is identified using which the graph is segmented into two graphs. This is iteratively applied to each graph to obtain the required segmentation.

3.1 Algorithm

1. Extract the features of the speech and represent it as $X = \{X_1, X_2, \ldots X_n\}$, where X_i is the ith feature vector of the given speech S.
2. Build a multigraph G = (V, E), consider each feature vector as a node and the relationship between the vectors as edge. Each edge is represented through a similarity measure between each node and graph is represented as a weight matrix (W) as follows,

$$w(i,j) = \begin{cases} e^{\frac{-\|X_i - X_j\|_2^2}{\sigma_X^2}} & \text{if } distance(X_i - X_j) < r \\ 0 & \text{Otherwise.} \end{cases} \tag{5}$$

3. Find the eigenvectors with the median eigenvalues of the system by representing it as a standard eigenvalue problem,

$$(D - W)x = \lambda Dx \tag{6}$$

where D a diagonal matrix with elements $d_i = \sum_j w(i,j)$, the weight of w_i. Let c_1, c_2, ... c_n be the sorted list of eigenvalues and E_1, E_2, ... E_n be their corresponding eigenvectors.
4. Starting from the eigenvector of second median eigenvalues, an eigenvector is selected for each level of segmentation and then an optimal cut is identified for the graph considered.
   (a) If E_i is the selected eigenvector, then 0, $mean(E_i)$ and $median(E_i)$ are considered as candidate values to bipartite the graph using the selected eigenvector.
   (b) The Ncut values for the candidate options in step 4(a) are calculated using (2). The candidate having a greater Ncut value is then considered as an optimal candidate.
5. The optimal candidate identified in step 4 is used to bipartite the graph. Then step 4 is repeated for each partition (subgraph) as graph until the required level of segmentation is achieved.

4 Experimental Results

The experiments were conducted on locally built speech corpus which includes the speech of 39 speakers in the age group between 17 and 45. Each speaker has spoken for about five minutes of speech that comprises of 45 sentences. This corpus as a whole contains about 195 min of speech. The speech used for this work is continuous speech spoken in Tamil language. A raw spoken sentence is initially filtered using the second order filter for noise removal, which is given as,

$$Y(i) = b_0 S(i) + b_1 S(i-1) + b_2 S(i-2) \qquad (7)$$

where S represents the input speech and Y, the filtered speech. The filtered speech is given as input to this algorithm which is first subjected to DWT feature extraction. Here, Daubachies wavelet (db2) is applied on the speech to extract the features at different frequency bands. The speech is decomposed to six levels and 6 Low frequency band (LFB) features are considered for further experiments. The feature vectors extracted so are then considered as nodes of the graph and the likelihood of the nodes as their edge weight as explained in the above algorithm to build a graph. The physical location of the nodes in the speech influences the existence of edges between nodes in the graph. The node distance (r) decides the existence of an edge. In this experiment, the distance factor (r) is set to 50. Figure 2 shows the weight matrix of a sample speech

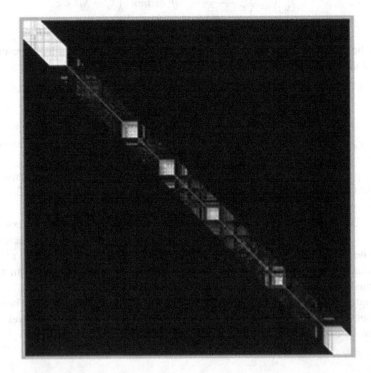

Fig. 2. Weight matrix of the graph G constructed from the feature vectors

'Thinai ennum sollukku ozhukkam enbathu porul'. The weight matrix shows the edges in the diagonal part of the matrix which is due to the restriction of the node distance on the edge existence.

The problem of segmentation is converted into a standard eigenvalue problem as represented in the step 3 of the algorithm and the eigenvalues and their corresponding eigenvectors are calculated. The selection of eigenvectors plays a crucial role in the segmentation accuracy. Experiments were performed by selecting eigenvectors corresponding to high, low and median eigenvalues respectively. The result of the experiments on two different speech samples spoken by two different individuals (1 female and 1 male) is shown in the Table 1, Fig. 4 and Table 2, Fig. 5 respectively. The study shows that choosing the eigenvectors with the median eigenvalues for successive levels of partitioning of the Graph produces better segmentation results. Figure 3 shows the sorted eigenvalues for the corresponding eigenvectors. The eigenvectors of median eigenvalues are considered for graph partitioning as it supports in better segmentation [13].

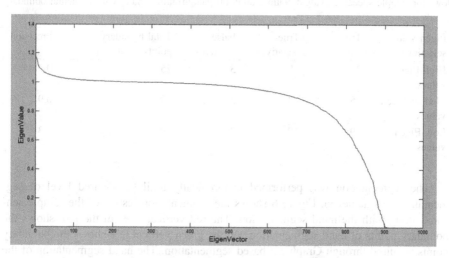

Fig. 3. Sorted eigenvalues for sample speech 1 'thinai ennum sollukku ozhukkam enbathu porul' (female)

Table 1. Result of choosing eigenvectors belonging to different eigenvalues (high, median and low) for sample speech 1 'thinai ennum sollukku ozhukkam enbathu porul' (female) with actual boundary points of 37

Eigen value selection	True positives	True negatives	False positives	Total boundary points identified	Precision
High Eigen values	30	4	8	38	0.8
Median Eigen values	**37**	0	11	48	1
Low Eigen values	23	14	7	30	0.6

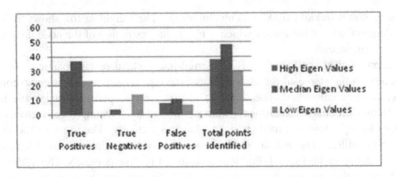

Fig. 4. Accuracy of segmentation based on the selection of eigenvalues for sample 1

Table 2. Result of choosing eigenvectors belonging to different eigenvalues (high, median and low) for sample speech 2 'vaigai Nathi mathuraiil paaigirrathu' (male) with 27 actual boundary points

Eigen value selection	True positives	True negatives	False positives	Total boundary points identified	Precision
High Eigen values	20	5	3	25	0.7
Median Eigen values	25	3	3	28	0.9
Low Eigen values	9	10	3	19	0.3

The segmentation was performed hierarchically until the desired level of segmentation was achieved. Figure 6 shows the segmentation results of the sample sentence along with the hand segmentation. The red vertical lines in the plot shows the boundary points of hand segmentation and the green vertical line shows the boundary points acquired through Graph cut based segmentation. The hand segmentation of the sample speech had 37 boundary points. The proposed graph cut based segmentation algorithm resulted with 37 true positives (boundary points that were detected correctly), 0 true negatives (undetected boundary points) and 11 were false positives (wrongly detected boundary points), totally 48 points for sample 1: 'thinai ennum sollukku ozhukkam enbathu porul'. And the result of applying the algorithm on sample 2: 'vaigai Nathi mathuraiil paaigirrathu' tured out with 25 true positives, 3 true negatives and 3 false positives for which the hand segmentation had 27 actual boundary points. The Fig. 7 and Table 3 shows the accuracy measures precision, recall and F-measure for the experiments conducted on Blind Segmentation using Non-linear filters (BSNLF) [2], Non-uniform segmentation using DWT (NUSDWT) [6] and the proposed Graph cut based segmentation methods. It shows the proposed method performs better in terms of precision, recall and F-measure which is 0.9259, 0.8928 and 0.9090 respectively.

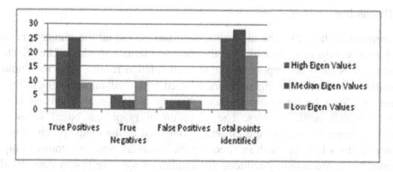

Fig. 5. Accuracy of segmentation based on the selection of eigenvalues for sample 2

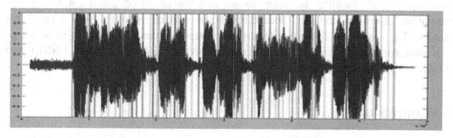

Fig. 6. The sample speech 'thinai ennum sollukku ozhukkam enbathu porul' showing the hand segmentation points (red) and graph cut based segmentation points (green) (Color figure online)

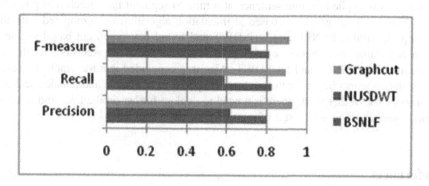

Fig. 7. Precision, recall and F-measure of graph cut, NUSDWT and BSNLF algorithms

Table 3. Comparing the accuracy of graph cut based segmentation with BSNLF and NUSDWT

Algorithm	Precision	Recall	F-measure
BSNLF	0.7977	0.8189	0.8067
NUSDWT	0.6156	0.5860	0.7172
Graph cut	0.9259	0.8928	0.9090

4.1 Discussion

The experiments show that the algorithm identifies almost all the phonetic boundary points for the sentences spoken by both the male and female speakers. The selection of the eigenvectors plays an important role in segmentation results. Also the number of true positives increases with eigenvectors selected with median eigenvalues. The false positives are from longer phonemes and missing points (true negatives) are from shorter phonemes or diphthongs. The precision of graph cut based segmentation algorithm is better for female speech when compared to the male speech. This algorithm uses a statistical approach in choosing the candidate splitting points to bipartite the graph. So, it theoretically reduces the number of splitting points considered and thus the time taken to identify the optimal cut. This algorithm blindly works on the speech given as input without referring any model, which infers it is independent of the vocabulary used. Finally, the study shows that the proposed graph cut based segmentation algorithm performs acceptably better and is irrelevant of the speaker's gender. The results of this algorithm show better segmentation results in terms of precision, recall and F-measure when compared to the existing algorithms BSNLF and NUSDWT.

5 Conclusion

In this work, the graph cut based segmentation which has shown its success in image segmentation was employed on the speech data. The experiments were conducted on the speech corpus which includes of both male and female speech. The segmentation algorithm was applied on one sentence at a time to segment the speech into phonetic units. The results of graph cut based segmentation algorithm was compared with the existing algorithms BSNLF and NUSDWT and found that graph cut based segmentation algorithm show better segmentation results in terms of precision, recall and F-measure than BSNLF and NUSDWT. Also it is concluded that the graph cut based segmentation works better for the continuous speech of both male and female speakers. As a scope for further work, there is a need to explore further on the false and missing boundaries identified as a result of continuous speech segmentation.

References

1. Juneja, A., Espy-Wilson, C.: Segmentation of continuous speech using acoustic-phonetic parameters and statistical learning. In: Proceedings of 9th International Conference on Neural Information Processing, vol. 2, pp. 726–730, November 2002
2. Räsänen, O., Laine, U.K., Altosaar, T.: Blind segmentation of speech using non-linear filtering methods. In: Ipsic, I. (ed.) Speech Technologies, chap. 5. InTech Open Access, June 2011. ISBN 978-953-307-996-7
3. Cosi, P.: SLAM: a PC-based multi-level segmentation tool. In: Rubio Ayuso, A.J., López Soler, J.M. (eds.) Speech Recognition and Coding. NATO ASI Series, vol. 147, pp. 124–127. Springer, Heidelberg (1995)

4. Wickerhauser, V.: Proceedings of the Third International Conference on Wavelet Analysis and Its Applications (WAA), Chongqing, PR China. World Scientific, 29–31 May 2003

5. Tan, B.T., Lang, R., Schroder, H., Spray, A., Dermody, P.: Applying wavelet analysis to speech segmentation and classification. In: SPIE's International Symposium on Optical Engineering and Photonics in Aerospace Sensing, pp. 750–761. International Society for Optics and Photonics, March 1994

6. Ziółko, M., Gałka, J., Drwiega, T.: Wavelet transform in speech segmentation. In: Fitt, A.D., Norbury, J., Ockendon, H., Wilson, E. (eds.) Progress in Industrial Mathematics at ECMI 2008, pp. 1073–1078. Springer, Heidelberg (2010)

7. Ziółko, B., Manandhar, S., Wilson, R., Ziółko, M.: Phoneme segmentation based on wavelet spectra analysis. Arch. Acoust. 36(1), 29–47 (2011)

8. Sarada, G.L., Lakshmi, A., Murthy, H.A., Nagarajan, T.: Automatic transcription of continuous speech into syllable-like units for Indian languages. Sadhana 34(2), 221–233 (2009)

9. Jayasankar, T., Thangarajan, R., Selvi, J.A.V.: Automatic continuous speech segmentation to improve Tamil text-to-speech synthesis. Int. J. Comput. Appl. 25(1), 31–36 (2011)

10. Nagarajan, T., Murthy, H.A.: Subband-based group delay segmentation of spontaneous speech into syllable-like units. EURASIP J. Adv. Signal Process. 2004, 1–12 (2004)

11. Shi, J., Malik, J.: Normalized cuts and image segmentation. IEEE Trans. Pattern Anal. Mach. Intell. 22(8), 888–905 (2000)

12. Bleyer, M., Gelautz, M.: Graph-cut-based stereo matching using image segmentation with symmetrical treatment of occlusions. Signal Process. Image Commun. 22(2), 127–143 (2007)

13. Xiang, T., Gong, S.: Spectral clustering with eigenvector selection. Pattern Recogn. 41(3), 1012–1029 (2008)

14. Wu, Z., Leahy, R.: An optimal graph theoretic approach to data clustering: theory and its application to image segmentation. IEEE Trans. Pattern Anal. Mach. Intell. 15(11), 1101–1113 (1993)

15. Stan, A., Mamiya, Y., Yamagishi, J., Bell, P., Watts, O., Clark, R.A.J., King, S.: ALISA: an automatic lightly supervised speech segmentation and alignment tool. Comput. Speech Lang. 35, 116–133 (2016)

Segmentation of Retinal Blood Vessels Using Pulse Coupled Neural Network to Delineate Diabetic Retinopathy

T. Jemima Jebaseeli$^{(\boxtimes)}$, D. Sujitha Juliet, and C. Anand Devadurai

Department of Computer Sciences Technology,
Karunya University, Coimbatore, India
{jemima_jeba, sujitha, ananddevadurai}@karunya.edu

Abstract. Diabetic Retinopathy (DR) is the root cause for retinal blood vessel damages among the diabetic patients. If it is not identified and treated earlier, at the later stage it leads to 100% vision loss. Thus there is a need of a system to identify the early stage of DR, so that it can be treated according to ETDRS (Early Treatment Diabetic Retinopathy Study). The proposed Pulse Coupled Neural Network (PCNN) model segments the retinal blood vessels from the depigmented fundus images and provides the structure of the retinal blood vessels. This segmented blood vessel map helps the ophthalmologist to identify the severity level of the blood vessel damages and to treat the early Diabetic Retinopathy among different age group populations. The proposed PCNN model is applied over the DRIVE database and the results are compared with various supervised and unsupervised segmentation approaches. The proposed method improves the accuracy in detecting the tiny blood vessels in the depigmented fundus images than other existing methods. This system increases the number of true positives; true negatives and reduces the false positives, false negatives while compared with the ground truth images. The Specificity of the proposed system over DRIVE database is 99.31%, Sensitivity is 67.54% and Accuracy is 97.23%. The resultant image of the segmented blood vessels can be used for further diagnosis and to measure the severity level of DR.

Keywords: Retinal blood vessel · Diabetic Retinopathy (DR) · PCNN · Pulse Coupled Neural Network · Segmentation

1 Introduction

The Diabetic Retinopathy (DR) which affects the retina of the working age group population and people with Type 2 diabetes. It is due to high blood pressure and high glucose level in the body [17]. There are chances for pregnant women to be affected by the DR during their pregnancy period. Even small children can have the possibility of occurrence of this disease. At later stage it leads to vision impairment and vision loss [26].

The damage of retinal blood vessel causes leakages of proteins and lipids into the retina while the nerves continue to carry the blood into the retina. The proteins and lipid deposit in the retina blocks the flow of blood and makes the outer layer of the retina to thicken [31].

© Springer Nature Singapore Pte Ltd. 2016
S. Subramanian et al. (Eds.): CSI 2016, CCIS 679, pp. 268–285, 2016.
DOI: 10.1007/978-981-10-3274-5_22

While rubbing the eye, the outer layer of the retina burst out and causes bleeding. Due to this lot of exudates, micro aneurysms, cotton wool spots, hemorrhages are spread inside the retina [13]. If it is identified very early, then one can be rescued from vision loss [15, 23]. Thus there comes a need for a system to predict the retinal vessel damages. The retinal blood vessel segmentation techniques have been used to identify the retinal vessel damages.

There are different segmentation techniques proposed by various researches, while all these methods work only on the images without any pigmentation [21, 27, 28]. Still there are problems in identifying the vascular tree [12] without any discontinuities in identifying the tiny blood vessels [1, 14, 16, 22]. The Otsu [35] method produces good segmentation results. However with complex multimodal images, the segmentation effect of this algorithm is not ideal. This method only takes the background of the image and the goal is to find the difference between the two classes of the objects.

The proposed Pulse Coupled Neural Network (PCNN) model segment the retinal blood vessel on depigmented images, it detects the vessel at their cross boundaries if it is suppressed with any other abnormalities also capable of detecting the tiny blood vessels. This process can be applied for any retinal fundus image databases.

2 Methods

The PCNN (Pulse Coupled Neural Network) is an unsupervised model which can operate without any training. The significant advantage of the network is that it is self-organized [7, 20]. This model can be used for image segmentation, image thinning, motion detection, pattern recognition, face detection, image de-noising, image enhancement, image fusion.

PCNN is the robust algorithm which is anti-noise against translation, scale, rotation of the input image patterns. The neurons in the network have the ability of responding to the stimuli. The PCNN firing matrix includes the geometry of the image to identify and remove the Gaussian noise and impulse noise [4, 8].

There are various methods that have been proposed to improve the PCNN parameters and to reduce the iteration time in order to achieve good segmentation results [1, 18, 19]. There is no method to automatically locate all the necessary parameters of PCNN [6] in order to accomplish the precise image segmentation. The parameters used in PCNN has been reduced by the simplified PCNN model with immune algorithm [5], Gray Scale Iteration Threshold [2, 3] Cross entropy, Shannon entropy, information entropy, Otsu method [9, 10], unit linking, Bayes clustering method, Nearest Neighbor Clustering Algorithm [11] etc.

The proposed PCNN model in this paper is proficiently determining the best choice of segments matching the target. This model identifies the retinal blood vessels from depigmentation. Hence it is capable to distinguishing the tiny blood vessels without any discontinuities.

3 Retinal Blood Vessel Segmentation Using PCNN

In this proposed method the retinal blood vessels are segmented using PCNN model within 5 iterations as shown in the Fig. 10, which is the less iteration level obtained in the depigmented retinal images of the DRIVE database [29] could be used for the classification of DR as mild, moderate and severe based on the formation of the vessel tree.

3.1 Preprocessing

Retinal images have luminosity deviations; they are noisy and have poor contrast. Thus pre-processing is required for the input images as shown in Fig. 1. The retinal images have three different channels; where the green channel is extracted for processing. In fundus image, the intensity of the vessel pixels is darker than the background. The contrast of the vessel pixel is more likely available only in the green channel of the RGB image. Hence the intensity of each pixel in the green channel image is taken for further processing. The green channel of the retinal image is preprocessed by the Adaptive Histogram Equalizer.

3.1.1 Adaptive Histogram Equalization (AHE)

The contrast of the image is enhanced by transforming the values on small regions called tiles. The enhancement is done on the contrast of each data region rather than the entire image. The output of the histogram region will match approximately to the specified histogram. The algorithm of Adaptive Histogram Equalization is as follows:

```
for each I(x,y)
do
{
    rank = 0
    for each I(i,j) in tile region of I(x,y)
    do
    {
        if I(x,y) > I( i,j) then
        rank = rank+1
    }
output_I(x,y)=rank*max_intensity(pixels in tile region)
}
```

where I(x,y) is the input image, I (i,j) is the tile region of the image. The tile size is automatically fixed based on the retinal vascular local geometry. The intensity of each pixel is transformed based on the histogram of the surrounding pixel with maximum intensity. The contrast enhanced image is shown in Fig. 9. Hence this process yields a high contrast fundus image which increases the accuracy in the consecutive segmentation results.

3.2   Processing

The contrast enhanced image is segmented by means of PCNN model as shown in Fig. 10. The neural networks are typically organized as layers. These Layers are made up of a number of interconnected nodes which carry out an activation function as shown in Fig. 2. In Fig. 3, the input patterns are presented to the network via the input layer, which communicates with one or more hidden layers where the actual processing is taking place via a system of weighted connections.

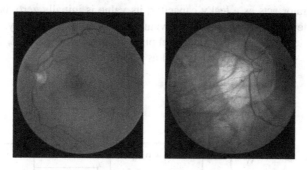

Fig. 1. The original color retinal image (DRIVE) (Color figure online)

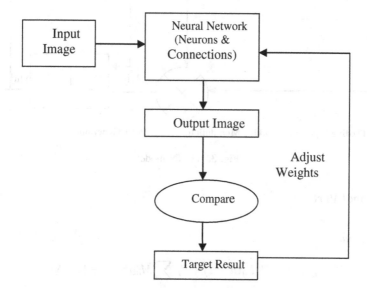

Fig. 2. Function of a neural network

The hidden layers are linked with the output layer. The each generated output would be given back to the network through a feedback layer to choose the best choice of segments matched with the target.

272   T. Jemima Jebaseeli et al.

3.2.1 The Normal PCNN Model

The PCNN model has three Fields: Feeding input Field, Linking Modulation, Pulse Generator. The decision to fire a neuron depends on its eight neighbors. The neuron in the neural network corresponds to an input image pixel of the input pattern. These neuron operations are described by the iterative equations. To compute F_{ij} the feeding input value of a neuron, where it initially has the pixel intensity value in i^{th}, j^{th} position. Each neuron is connected or linked to another neuron with synaptic weight W. The linking force depends on β. It calculates the L_{ij}. U_{ij} is the dynamic threshold value for a pixel.

Pulse generator acts as a leaky capacitor; its value is initially zero, because of the intensity of the neighboring pixel, its value increases exponentially. At dynamic threshold E, it decays; for non-blood vessel pixel it cannot increase its E_{ij} larger. Hence the pixel $U_{ij}(n) < E_{ij}(n-1)$ is considered as non-blood vessel.

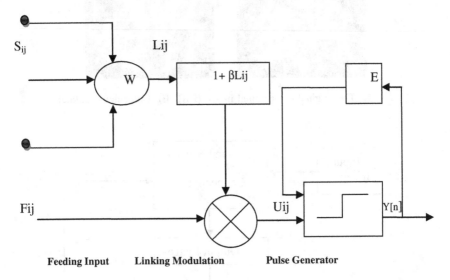

Fig. 3. PCNN model

3.2.1.1 Input Part

Feeding Input:

$$F_{ij}(n) = e^{-\alpha_F} F_{ij}(n-1) + V_F \sum_{kl} M_{ijkl} Y_{kl}(n-1) + S_{ij} \tag{1}$$

Where (i,j) is the position of the neuron in the network to identify the pixel which is present in the input image. The feeding input takes the normalized images obtained through AHE. If the input image is of 128×128 dimensions, then (i,j) will be between $(1,1)$ to $(128,128)$. (k,l) is the position of the surrounding neurons in the image. n is the iterative step number. S_{ij} is the gray level of the input image pixel. F_{ij} is the feeding input of the network. α_F, α_L, α_E are the attenuation time constants. V_F, V_L, V_E are the voltage potential of the feeding signal, linking signal and dynamic threshold.

Linking Input:

$$L_{ij}(n) = e^{-\alpha_L} L_{ij}(n-1) + V_L \sum_{kl} W_{ijkl} Y_{kl}(n-1) \tag{2}$$

L_{ij} is the linking input of the network. M, W is the synaptic weights, their values depends on the surrounding linking field neurons.

3.2.1.2 Linking Part

Internal Activity of the Neuron:

$$U_{ij}(n) = F_{ij}(n) \left(1 + \beta L_{ij}(n)\right) \tag{3}$$

U_{ij} is the internal activities of the network. β is the linking coefficient constant.

3.2.1.3 Pulse Generator

Output:

$$Y_{ij}(n) = \begin{cases} 1 & U_{ij}(n) > E_{ij}(n-1) \\ 0 & Other \end{cases} \tag{4}$$

E_{ij} is to set the dynamic threshold value in the network. Y_{ij} is the output neuron and its binary value will decide the status of the neuron as shown in Fig. 4.

Threshold Potential:

$$E_{ij}(n) = e^{-\alpha_E} E_{ij}(n-1) + V_E Y_{ij}(n) \tag{5}$$

The output neuron Y_{ij} contains the regional information, edge information and the features of the binary image according to Fig. 3. If Y_{ij} is equal to one then the neuron is activated.

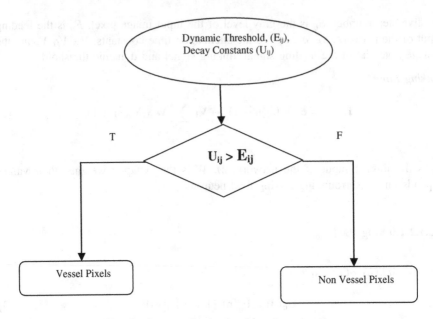

Fig. 4. Feature selection by the active neurons.

3.2.1.4 PCNN Output

The PCNN model produces the feature vector as the output which is used to identify the blood vessels from each active neuron in the network. The feature vector can be obtained by summing up the pixel values in the output binary image.

$$\text{The feature vector } G(n) = \sum_{ij} Y_{ij}(n) \tag{6}$$

The length of the feature vector is the total number of iteration steps in the PCNN model. If the iteration is n times then the feature vector has the length of n elements. The quality of the output image depends on the length of the feature vector. The benefit of PCNN model is that the generated feature vector is unique for the image which is used to recognize the vessel pixels accurately and it is free from noise. The feature vector is invariant against the geometrical changes in the image.

In Fig. 5, the input images are taken from the DRIVE dataset as shown in Fig. 1, and given as an input to the system for preprocessing, whereas the green channel of the retinal image has been taken and the pixels are enhanced by AHE. The enhanced image is shown in Fig. 9 will be taken into processing, where the PCNN model segment only the blood vessels from the background and classify them from non- blood vessel pixels as shown in Fig. 10. This is obtained through the Algorithm 1. The morphological thinning processes darkens up the extracted vessels. Then, the PCNN method is compared with the predefined quality measures of the ground truth image, through which the algorithms performance is compared to show its successive retrieval of

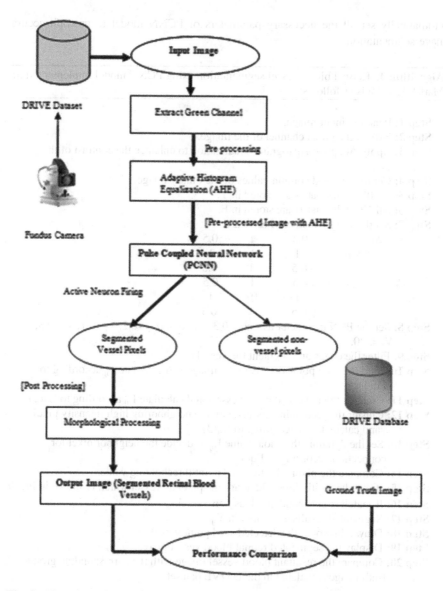

Fig. 5. The schematic diagram of the proposed retinal blood vessel segmentation method.

vessels. According to Fig. 6, the vessel pixels are highlighted as (∗) and the non-vessel pixels are highlighted as (.) which depicts the differences among the active neurons and non-active neurons.

In Fig. 7, the active neurons started firing with the seed point pixels which has high intensity value. Each active neuron will compare their decay value with the dynamic threshold value of the seed point vessel pixels. The Weight matrix will link this active neuron with the seed point vessel. Currently, no methods have been designed to automatically set all the necessary parameters of PCNN model to triumph accurate image segmentation.

Algorithm 1. Retinal blood vessel segmentation using PCNN model implemented in Matlab R2010a is as follows:

Step 1: Read the input image.
Step 2: Extract the green channel of the image.
Step 3: Apply Adaptive Histogram Equalization to enhance the contrast of the image.
Step 4: Get the row and column values of the binary image.
Step 5: All 0's in the matrix are stored in Y.
Step 6: All 1's in the matrix are stored in E.
Step 7: Set the weight matrix M & W as,

$$
M= \begin{matrix} 0.5 & 1 & 0.5 \\ 1 & 0 & 1 \\ 0.5 & 1 & 0.5 \end{matrix}
$$

$$
W= \begin{matrix} 0.5 & 1 & 0.5 \\ 1 & 0 & 1 \\ 0.5 & 1 & 0.5 \end{matrix}
$$

Step 8: Set the PCN parameter as, $\alpha_F=0.3$, $\alpha_L=0.5$, $\alpha_E=0.2$, $V_F=0.1$, $V_L=0.3$, $V_E=220$.
Step 9: Filter the edges and segment the vessels.
Step 10: Find the seed point vessels in the image and calculate F_{ij} according to Eq(1).
Step 11: Connect the next seed point vessels and calculate L_{ij} according to Eq(2).
Step 12: Cluster the seed point vessels with the neighboring high intensity pixels and calculate U_{ij} according to Eq(3).
Step 13: Set the dynamic threshold value E_{ij} to decide the neighbor pixel for connection according to Eq(4).
Step 14: Compare the pixel with the seed point thresholding value.
Step 15: If matched with threshold, mark the pixel as vessel which is shown in Fig. 7.
Step 16: If match is not found, mark as non-vessel.
Step 17: Segmented results are stored in Y_{ij}.
Step 18: Draw a line over the detected seed point vessel.
Step 19: Display the segmented retinal blood vessel image.
Step 20: Compare the resultant blood vessel image with the corresponding ground truth images available in the DRIVE dataset.

In Fig. 8, each pixels fitness value will be calculated and it will be compared with the neighboring pixels fitness value, if it is high then the pixel will be connected with the nearest neighbor pixels from the seed point vessel pixels. Thus the centerline of the vessel tree structure is formed by connecting all the tiny vessel pixels by comparing its fitness value with the dynamic threshold. The dynamic threshold shows the pixels with high intensity value; thus all the vessels will be clustered together with its nearest neighbor. Thus it can omit all other non-vessel pixels with lowest fitness value than the seed point vessel pixels.

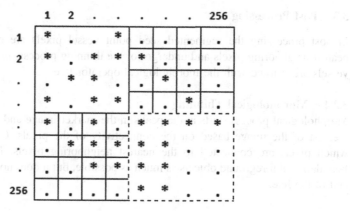

Fig. 6. The highlighted vessel and non-vessel pixels of 8 nearest neighbor pixels of the input fundus image.

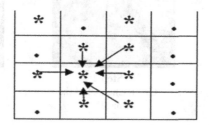

Fig. 7. Firing of neurons towards active neurons

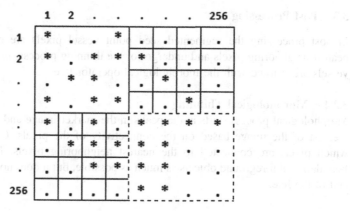

Fig. 8. The highlighted vessel pixels of 8 nearest neighbor pixels of the input fundus image.

3.3   Post Processing

In post processing the segmented seed point vessel pixels are connected with the nearest neighboring pixels and undergo for the thinning process, in which the detected vessels are marked with the morphological operator line.

3.3.1   Morphological Thinning

Morphological process starts at the peaks in the marker image and spreads throughout the rest of the image based on the connectivity of the pixels. Connectivity defines which pixels are connected to the nearest neighboring pixels. It erodes away the boundaries of foreground objects as much as possible, but it does not affect pixels at the end of the line.

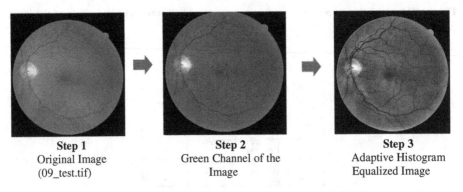

Step 1	Step 2	Step 3
Original Image	Green Channel of the	Adaptive Histogram
(09_test.tif)	Image	Equalized Image

Fig. 9. Preprocessing results of the fundus image (DRIVE).

4   Results

The results are obtained for the 40 images available in the DRIVE database. The obtained vascular tree structure of the input images are evaluated with the ground truth blood vessel image map and it shows that more accuracy is found in the proposed retinal vessel segmentation method.

The depigmentation found in the image at the vessel points are removed automatically by the PCNN model and enable the system to track all the tiny vessels hidden by this depigmentation. The vessels pixels connected with its nearest neighbors are clustered with the next 8-nearest neighboring vessel pixels. These are clustered together to form the structure of the vessel map.

The vessel map connects the outer vessels with the inner vessel line structures. The seed point vessel pixels connect all the interlinked vessel pixels together. Thus this seed point vessel clusters connects all other clustered vessel structures to form the complete vessel map.

The vessels on the boundary can be tracked and identified by their fitness value. If their fitness values are matched with the threshold value of the seed point vessel pixel, then they are identified as the vessel pixels.

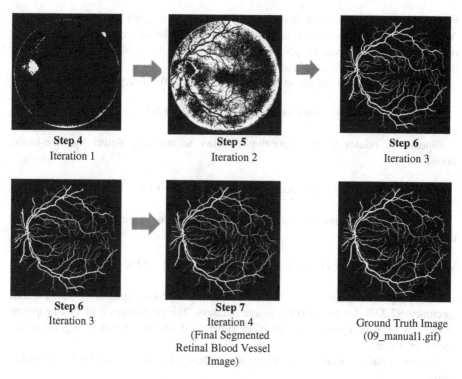

Fig. 10. Function of PCNN over retinal image (DRIVE)

The overlapping boundary vessels are also identified and they are connected with their corresponding vessel line. Finally, all other non-vessel pixels are omitted to form the vessel structure.

The 40 images of DRIVE dataset is considered for testing. Table 1 shows the generated sample results for the first ten images. The background pixels are masked while segmenting the vessel pixels. If the pixel location of ground truth image's where the vessel pixel is present and which is incorrectly identified as non-vessel pixel in the segmented image then it is marked as FP. If the pixel location of ground truth image's where the non-vessel pixel is present, then it is incorrectly identified as vessel pixel in the segmented image which is marked as FN. The same pixel location of both ground truth image and the segmented image contains vessel pixel then it is marked as TP. TN is high since, the same pixel location of both ground truth image and the segmented output image does not contain vessel pixels. The overall performance of this proposed algorithm is better than all other existing methods.

4.1 Performance Analysis

The parameters used for analyzing the performance of the resultant blood vessel segmentation images are True Positive (TP), True Negative (TN), False Positive (FP),

False Negative (FN), False Positive Rate (FPR), Sensitivity (SP), Specificity (SPE) and Accuracy (ACC). Where TP is the correctly identified blood vessel, FP is the incorrectly identified blood vessel, TN is the correctly rejected non blood vessel and FN is the incorrectly identified non blood vessel.

Sensitivity refers to the algorithm's ability to correctly detect the blood vessels.

$$\text{Sensitivity (SP)} = \text{TP}/(\text{TP}+\text{FN})$$

Specificity relates to the algorithm's ability to correctly detect the non-blood vessels.

$$\text{Specificity (SPE)} = \text{TN}/(\text{TN}+\text{FP})$$

Accuracy is the degree of closeness of measurements of a quantity to that quantity's true value.

$$\text{Accuracy (ACC)} = \text{TP}+\text{TN}/(\text{TP}+\text{FN})+(\text{TN}+\text{FP})$$

The proposed system has achieved Sensitivity **67.54%**, Specificity **99.31%** and Accuracy **97.23%** for the DRIVE image datasets. The performance of this proposed model has been compared with all other supervised and unsupervised algorithms as shown in Table 2.

The quantitative evaluation of the contrast enhanced image is done by the following metrics.

(1) The Image Contrast (I_C) between the retinal vessels (Y) and the background (G) is measured by,

$$I_C = \frac{I_Y - I_G}{I_Y + I_G} \tag{7}$$

where I_Y is the average gray values of the retinal vessels and I_G is the average gray values of the background.

(2) The Contrast Rate (I_{CR}) between the Enhanced Image (I_{CE}) and the Source Image (I_{SE}) is measured by,

$$I_{CR} = \frac{I_{CE}}{I_{SE}} \tag{8}$$

In this experiment the value of I_C and I_{CR} is larger, so it highly differentiates the retinal vessels from the background. The DRIVE dataset contains 40 images. In which 7 of the images are pathological images. These pathological images are infected diseased images, which contains microaneurysm, hemorrhages, exudates, cotton wool spots. These particles cause more depigmentation in the image. But the PCNN model identifies the blood vessels among these 7 pathological images with high accuracy of 97.23%. This proves that the algorithm is efficient to detect the blood vessels in spite of depigmentation among the images.

Table 1. Generated segmentation results by the proposed method on DRIVE dataset for 1–10 sample images.

Source image	TP	TN	FP	FN	FPR	Sensitivity %	Specificity %	Accuracy %
01_test.tif	489	264414	11338	6726	0.041116656	67.77	99.81	93.61
02_test.tif	383	261578	11286	5649	0.041361264	63.49	99.85	93.92
03_test.tif	3209	222370	7897	46587	0.034294971	64.44	98.57	80.54
04_test.tif	2028	236747	9300	35022	0.037797657	54.73	99.15	84.34
05_test.tif	3507	221608	8466	47823	0.036796857	68.32	98.44	79.99
06_test.tif	1617	245396	11046	21011	0.043074067	71.46	99.34	88.51
07_test.tif	2268	234271	8571	39972	0.035294554	53.69	99.04	82.97
08_test.tif	1739	241636	9478	30092	0.037743814	54.63	99.28	86.01
09_test.tif	217	272584	10769	2937	0.038005597	68.80	99.92	95.21
10_test.tif	2989	220905	8356	53809	0.036447542	52.62	98.66	78.26

The proposed method is able to detect all the tiny vessels without any deviation. The algorithm is sensitive to identify the non-vessel pixels. Hence it's able to classify the non-vessel pixels from the actual vessel pixels. The sensitivity of the proposed algorithm is 67.54%. The algorithm shows that it specifies the vessel pixels up to 99.31% as shown in Table 2. Thus the proposed method is better in identifying the tiny vessels from the depigmented images among the pathological images of DRIVE datasets.

5 Discussions

It is important in the clinical analysis to accurately identify the retinal blood vessels for analyzing the severity level of Diabetic Retinopathy. Many algorithms are not able to identify the retinal blood vessels from the depigmented pathological retinal images.

There are algorithms which are misclassifying the non-vessel pixels as vessel pixels and it leads to false identification of vessels and increases the False Negative Rate (FNR) and decreases the False Positive Rate (FPR) in classifying the vessel pixels as non-vessel pixels.

Some algorithms are difficult in identifying the vessel pixels in the cross boundary section. Due to depigmentation the paler exudates causes the system to fail in misclassification of vessels pixels. Table 2 shows that the performance of the proposed algorithm is better than all other methods which are having lower performance on DRIVE datasets.

In the Compensation Factor Method [22], the presence of noise which degrades the image and enhancement may pick up some more additional noise which leads to false vessel detection. Hence, extracting the complete vascular tree, there are discontinuities in the vessel branches. In severely damaged images the performance is low. The vessels inside the disk have weak connections with the neighborhood pixels.

In Adaboost Algorithm [23], due to retinal occlusions, the images are classified as abnormal, even if the patients do not suffer from DR. In Level Set and Region Growing [24], method some thin vessels are missed in segmentation, since their grey values are

Table 2. Performance comparisons of different techniques for retinal vessel segmentation on the DRIVE database.

Method type	Name of the author and year of publication	Name of the method	Sensitivity (%)	Specificity (%)	Accuracy (%)
Supervised	Salazar et al. [22]	Compensation factor method	84.64	-	-
	Zhao et al. [24]	Level set and region growing	73.54	97.89	94.77
	Welikala et al. [30]	Standard line operator	87.93	94.40	-
	Ramlugun et al. [34]	Matched filter based approach	64.13	97.67	93.10
Unsupervised	Franklin et al. [25]	Back propagation feed forward network	-	-	95.03
	Marin et al. [33]	Back propagation multilayer neural network	70.67	98.01	94.52
	Mookiah et al. [32]	Probabilistic neural network	96.27	96.08	96.15
The proposed method		Pulse coupled neural network	67.54	99.31	97.23

similar to background and their width span is narrow in range. In the Standard Line Operator method [30], the evaluation of new vessel pixels cannot be distinguished from non-vessel edges in the vessel map.

In the Matched Filter Based Approach [34], the preprocessing technique fusing the vessels with the background and it is difficult to segment them out because of the low contrast with the background. It decreases the vessel diameter along the vascular length.

The Back propagation Neural Networks are applied to the image pixels without any prior feature extraction. Each pixel is classified as raw pixel values in a square window centered on it are fed to the network as input pattern. The proposed method can segment the retinal blood vessels including the normal and abnormal images. The PCNN model is an unsupervised neural network and it doesn't need any input patterns. The network can train itself and the feature vectors are generated automatically for identifying the vessel pixels. The PCNN is robust and anti-noise against translation, scale and rotation of the input process.

The connection coefficients, decay time constant and weighing factors are to be set in advance. The result quality depends on loop iterations. This proposed PCNN network can segment the retinal blood vessels within 4 iterations. The algorithm performs vessel segmentation only in the pixels and its 8 neighborhood pixels and it avoids processing of all other pixels in the image.

PCNN is better than median filter, wiener filter, Lee filter, anisotropic diffusion based filters. The PCNN firing matrix includes the geometry structures of the image to detect and remove noise. PCNN removes noise without affecting the tiny blood vessel regions.

The PCNN model has overall performance improvement on DRIVE dataset with Sensitivity 67.54%, Specificity 99.31%, Accuracy 97.23% over normal and abnormal images. The system doesn't fail in identifying the tiny blood vessels which are hidden by the depigmentation caused by the infected particles.

6 Conclusions

The proposed method can reduce the work of the ophthalmologists in analyzing the blood vessels of the patients with Diabetic Retinopathy by analyzing the segmented vessel structure. The experimental results are obtained for all the 40 images of the DRIVE Database. The results prove that the proposed Adaptive Histogram Equalization (AHE) with Pulse Coupled Neural Network (PCNN) model can detect all the tiny blood vessels precisely. The tiny vessels are distinguished without any discontinuities. The PCNN model eradicates the noise presents in the depigmented retinal images automatically. The parameters of PCNN model can further improved by means of including the optimization techniques. Our further research is on analyzing the vision of the children who are spending lot of time on electronic gadgets leads to sudden vision loss. Also, the infant's retinal vessels may grow out of the vessel boundary that is to be identified early for treatment. Hence the proposed retinal vessel segmentation techniques can be applied on similar cases of datasets. The tortuosity of the vessel is also an important feature needed for the DR system while analyzing the retina that will be considered in our next work.

References

1. Wang, S., Xiao, Z., Wu, J., Geng, L., Zhang, F., Xi, J.: Fundus blood vessels detection based on pulse coupled neural network. Int. J. Digit. Content Technol. Appl. **6**(15), 467–474 (2012)
2. Ma, H.-R., Cheng, X.-W.: Automatic image segmentation with PCNN algorithm based on grayscale correlation. Int. J. Signal Process. Image Process. Pattern Recogn. **7**(5), 249–258 (2014)
3. Li, H., Lei, G., Yufeng, Z., Shi, X., Jianhua, C.: A novel method for grayscale image segmentation by using GIT-PCANN. Int. J. Inf. Technol. Comput. Sci. **5**, 12–18 (2011)
4. Subashini, M.M., Sahoo, S.K.: Pulse coupled neural networks and its applications. Expert Syst. Appl. **41**, 3965–3974 (2014)
5. Li, J., Zou, B., Ding, L., Gao, X.: Image segmentation with PCNN model and immune algorithm. J. Comput. **8**(9), 2429–2436 (2013)

284 T. Jemima Jebaseeli et al.

6. Xu, X., Ding, S., Shi, Z., Zhu, H., Zhao, Z.: Particle swarm optimization for automatic parameters determination of pulse coupled neural network. J. Comput. **6**(8), 1546–1553 (2011)

7. Cheng, D., Zhao, W., Tang, X., Liu, J.: Image segmentation based on pulse coupled neural network. In: Proceedings of the 11th Joint Conference on Information Sciences (2008)

8. Zhang, D., Mabu, S., Hirasawa, K.: Noise reduction using genetic algorithm based PCNN method. IEEE (2010)

9. Guerrout, E.H., Mahiou, R., Ait-Aoudia, S.: Hidden Markov random fields and swarm particles: a winning combination in image segmentation. IERI Procedia **10**, 19–24 (2014)

10. Raja, N.S.M., Sukanya, S.A., Nikita, Y.: Improved PSO based multi-level thresholding for cancer infected breast thermal images using Otsu. Procedia Comput. Sci. **48**, 524–529 (2015)

11. Salem, S.A., Salem, N.M., Nandi, A.K.: Segmentation of retinal blood vessels using a novel clustering algorithm (RACAL) with a partial supervision strategy. Med. Biol Eng. Comput. **45**, 261–273 (2007)

12. Hao, J.T., Li, M.L., Tang, F.L.: Adaptive segmentation of cerebrovascular tree in time-of-flight magnetic resonance angiography. Med. Biol. Eng. Comput. **46**, 75–83 (2008)

13. Khademi, A., Krishnan, S.: Shift-invariant discrete wavelet transform analysis for retinal image classification. Med. Biol. Eng. Comput. **45**, 1211–1222 (2007)

14. Dougherty, G., Johnson, M.J., Wiers, M.D.: Measurement of retinal vascular tortuosity and its application to retinal pathologies. Med. Biol. Eng. Comput. **48**, 87–95 (2010)

15. Fadzil, M.H.A., Izhar, L.I., Nugroho, H., Nugroho, H.A.: Analysis of retinal fundus images for grading of diabetic retinopathy severity. Med. Biol. Eng. Comput. **49**, 693–700 (2011)

16. Pereira, C., Goncalves, L., Ferreira, M.: Optic disc detection in color fundus images using ant colony optimization. Med. Biol. Eng. Comput. **51**, 295–303 (2013)

17. Zarkogianni, K., Nikita, K.S.: Special issue on emerging technologies for the management of diabetes mellitus. Med. Biol. Eng. Comput. **53**, 1255–1258 (2015)

18. Jiang, W., Zhou, H., Shen, Y., Liu, B., Fu, Z.: Image segmentation with pulse-coupled neural network and Canny operators. Comput. Electr. Eng. **46**, 528–538 (2015)

19. Berg, H., Olsson, R., Lindblad, T., Chilo, J.: Automatic design of pulse coupled neurons for image segmentation. Neurocomputing **71**, 1980–1993 (2008)

20. Xie, W., Li, Y., Ma, Y.: PCNN-based level set method of automatic mammographic image segmentation. Optik **127**, 1644–1650 (2016)

21. Agurto, C., Yu, H., Wigdahl, J., Pattichis, M., Nemeth, S., Barriga, E.S., Soliz, P.: A multiscale optimization approach to detect exudates in the macula. IEEE J. Biomed. Health Inform. **18**(4), 1328–1336 (2014)

22. Salazar-Gonzalez, A., Kaba, D., Li, Y., Liu, X.: Segmentation of the blood vessels and optic disc in retinal images. IEEE J. Biomed. Health Inform. **18**(6), 1874–1886 (2014)

23. Roychowdhury, S., Koozekanani, D.D., Parhi, K.K.: DREAM: diabetic retinopathy analysis using machine learning. IEEE J. Biomed. Health Inform. **18**(5), 1717–1728 (2014)

24. Zhao, Y.Q., Wang, X.H., Wang, X.F., Shih, F.Y.: Retinal vessels segmentation based on level set and region growing. Pattern Recogn. **47**, 2437–2446 (2014)

25. Franklin, S.W., Rajan, S.E.: Computerized screening of diabetic retinopathy employing blood vessel segmentation in retinal images. Biocybern. Biomed. Eng. **34**, 117–124 (2014)

26. Akram, M.U., Khalid, S., Tariq, A., Khan, S.A., Azam, F.: Detection and classification of retinal lesions for grading of diabetic retinopathy. Comput. Biol. Med. **45**, 161–171 (2014)

27. Sharma, P., Nirmala, S.R.: A system for grading diabetic maculapathy severity level. Netw. Model. Anal. Health Inform. Bioinform. **3**, 49 (2014)

28. Akram, M.U., Khalid, S., Khan, S.A.: Identification and classification of microaneurysms for early detection of diabetic retinopathy. Pattern Recogn. **46**, 107–116 (2013)

29. The DRIVE dataset. http://www.isi.uu.nl/Research/Databases/DRIVE/
30. Welikala, R.A., Dehmeshki, J., Hoppe, A., Tah, V., Mann, S., Williamson, T.H., Barman, S. A.: Automated detection of proliferative diabetic retinopathy using a modified line operator and dual classification. Comput. Methods Programs Biomed. **114**, 247–261 (2014)
31. Sopharak, A., Uyyanonvara, B., Barman, S.: Simple hybrid method for fine microaneurysm detection from non-dilated diabetic retinopathy retinal images. Comput. Med. Imaging Graph. **37**, 394–402 (2013)
32. Mookiah, M.R.K., Acharya, U.R., Martis, R.J., Chua, C.K., Lim, C.M., Ng, E.Y.K., Laude, A.: Evolutionary algorithm based classifier parameter tuning for automatic diabetic retinopathy grading: a hybrid feature extraction approach. Knowl. Based Syst. **39**, 9–22 (2013)
33. Marin, D., Gegundez-Arias, M.E., Suero, A., Bravo, J.M.: Obtaining optic disc center and pixel region by automatic thresholding methods on morphologically processed fundus images. Comput. Methods Programs Biomed. **118**, 173–185 (2015)
34. Ramlugun, G.S., Nagarajan, V.K., Chakraborty, C.: Small retinal vessels extraction towards proliferative diabetic retinopathy screening. Expert Syst. Appl. **39**, 1141–1146 (2012)
35. Yao, C., Chen, H.: Automated retinal blood vessels segmentation based on simplified PCNN and fast 2D-Otsu algorithm. J. Cent. South Univ. Technol. **16**, 0640–0646 (2009)

Assessment of Bone Mineral Health of Humans Based on X-Ray Images Using Inference

Geetha Ganapathi[1(✉)] and N. Venkatesh Kumar[2]

[1] Department of Applied Maths and Computational Sciences,
PSG College of Technology, Coimbatore 641 004, India
ngg@amc.psgtech.ac.in
[2] Department of Orthopedics, PSG Institute of Medical Sciences and Research,
Coimbatore 641 004, India

Abstract. Bones that provide the structural support of the body, are composed of many inorganic compounds and organic materials that all together can be used to determine the mineral density of the bone. The bone mineral density (BMD) is an index measure that is widely used as an indicator of the health of the bone. A densitometry study from dual X-ray absorptiometry (DEXA) system is a popularly used method to assess BMD. BMD values vary depending on race, age, gender and other health conditions. As DEXA is quite an expensive method and requires frequent calibration process to work properly, in this paper, we explore the possibility of developing an affordable and reliable system depending on single X-ray absorptiometry with the use of supervised learning methods. The methodology based on inference is tested on a data set consisting of spine and pelvis X-ray images of patients of varying ages between 10 and 90 years of PSG Hospitals, Coimbatore, India and the results proved to be an indicator of density of the bone.

Keywords: BMD · Classification · Supervised learning · Support vector machines · Generalized boosted regression model · Extreme gradient boosting model

1 Introduction

Bones are dynamic connective tissue that constitute the endoskeleton of vertebrates. They support and protect many organs of the body, store minerals, produce red and white blood cells. Bones have complex internal and external structure, come in different shapes, hard and strong in nature but lightweight and serve multiple functions. Bones together with muscles, ligaments and joints help in movement of various body parts. Bone is made up of bone matrix. It is composed primarily of inorganic hydroxyapatite and organic collagen. Bone is formed by the process of mineralization and calcification of this matrix around entrapped cells. Remodelling is the process of resorption followed by replacement with little change in shape and occurs throughout a person's life which tends to

© Springer Nature Singapore Pte Ltd. 2016
S. Subramanian et al. (Eds.): CSI 2016, CCIS 679, pp. 286–299, 2016.
DOI: 10.1007/978-981-10-3274-5_23

phase down with advancement of age. The purpose of remodelling is to regulate and maintain calcium homeostasis, repair micro-damaged bones and to shape the skeleton during growth.

Bone mineral density (BMD) is a measure of the amount of hydroxyapatite (Hap) and mineral salts per unit area in the bone that is used as an indicator of the bone's health. The mineral content in a bone sample is evaluated by direct evaluation through the atomic absorption spectroscopy and inductively coupled plasma method or by indirect method, which consists of single or dual X-ray diffraction and computer tomography to obtain the bone mineral density. The indirect method of measurement is painless, non-invasive and involves low radiation exposure over the spine or hip region. By using direct methods, it is possible to obtain quantitative information regarding the bone mineral components, whereas in indirect methods, the information is closely related to the optical absorption of the bone tissues. Nowadays, BMD is measured by dual-energy X-ray absorptiometry (DEXA). The DEXA system uses double X-ray intensities in order to obtain two images, one of which is taken at high energy, whereas the second one is taken at low energy. The difference between the two is normally used to calculate the BMD by means of a calibration function of the system itself. WHO defines BMD to be 2.5 standard deviations (SD) or more below the peak bone mass as a normal T score (above $-1\,SD$); between -1 and $-2.5\,SD$ defined as mildly reduced bone mineral density and below $-2.5\,SD$ as osteoporosis.

The average density of the bone is around 1000 to 1200 mg/cm^2. The density varies with age, race and gender. Bone density tests measure two conditions, osteopenia and osteoporosis. BMD would be greater than or equal to 833 mg/cm^2 for normal people. Osteopenia is a condition where BMD is lower than normal, like between 648 mg/cm^2 and 833 mg/cm^2 [2]. With further bone loss, osteopenia leads to Osteoporosis, a disease of the bones, where BMD is still reduced (less than 648 mg/cm^2), bone micro-architecture deteriorates and the amount and variety of proteins in bone are altered. Poor bone density relates to higher probability of fracture. Bones become weak and can break from a minor fall or even from simple actions like sneezing and twisting. People over the age of 50, people with smoking habits, drinking habits, long term use of steroid drugs, Vitamin D deficiency and women with early menopause are more susceptible for having less bone density. Women are more susceptible to get Osteoporosis due to the reasons: (1) Women tend to have thinner, smaller bones than men (2) Estrogen, a hormone that protects bones, decreases sharply when women reach menopause, which can cause bone loss. Fractures are the most dangerous aspect of osteoporosis. Acute and chronic pain is often attributed to fractures from osteoporosis and can lead to further disability and early morbidity and functional impairment. In essence, the daily activities of a person become a painful ordeal and overall quality of life declines. Reduced bone mass is therefore a useful predictor of increased fracture risk. Normal and Osteoporotic conditions of a bone are shown in Fig. 1.

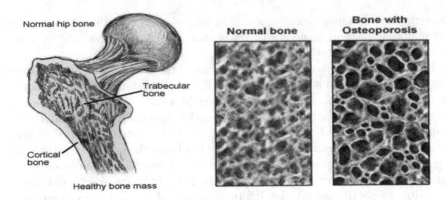

Fig. 1. Normal bone image: a. Trabecular bone b. Micro-architecture of normal and Osteoporosis affected bone

DEXA is the most widely used and most thoroughly studied bone density measurement technology. But it is quite an expensive method in developing countries like India. Moreover BMD values calculated depend on race, age and other conditions and not a constant for all people throughout the world. Hence, this paper explores the possibility of developing an affordable and reliable system to study bone mineral density using X-rays which is a very basic and conventional screening morality.

Section 2 discusses the related work. Section 3 discusses the proposed methodology used in this paper. Section 4 elaborates on image enhancement, feature extraction and bone mineral assessment using different classification methods. Section 5 reports on the experiments conducted and the results obtained. Finally, Sect. 6 concludes the work.

2 Related Work

DEXA is recognized as the acceptable method to measure BMD. DEXA scans can have errors between 5 and 8% due to calibration, positioning and other factors. X-rays are more cheaper and many researchers have shown that there exists relationship between plain radiographic patterns and three-dimensional trabecular architecture.

Luo et al. [12] have shown that the plain radiograph contains architectural information directly related to the underlying 3D structure based on the study of human calcaneal trabecular bone. Veenland [19] has assessed the suitability of different texture analysis methods for use in radiographs and also to select texture features that are able to quantify the changes in the radiographic trabecular pattern occurring in osteoporosis. Chappard et al. [3] showed that texture analysis by Euclidean and fractal methods provided significant differences in the two bones of a rat model submitted for disuse induced by Clostridium botulinum Toxin and found that such analysis on radiographs appears able to detect architectural differences in the trabecular architecture even when bone loss has not reached a value sufficient enough to induce changes appreciable by DEXA. Again, Chappard et al. [4] have shown that the trabecular characteristics were found to be highly correlated with texture parameters describing the X-ray image and suggested that the texture analysis of X-ray films might be a suitable approach to investigate the disorganization of bone in osteoporosis. X-ray films constitute a 2D projection of the trabecular architecture, the resulting image is texture and this may be an indicator of disease etiology.

Guggenbuhl et al. [9] in their study found a good correlation between texture analysis of X-ray radiographs and 3-D bone microarchitecture assessed by micro-CT of human iliac bone. Jimenez-Mendoza and Espinosa-Arbelaez [11] have determined the parameter equation that defines the ratio of the pixels in radiographs and BMD and demonstrated the methodology on Wistar rats' femur bones on a calibrated system.

All this suggest that the study of bone mineral health by direct means is tedious and harmful to the human body and so an indirect measure by the DEXA is an accepted methodology. With the technology yet to reach the common masses being hindered by cost and the BMD values available are only for the white race, the study of BMD using X-ray images is a boon for the developing nations like India.

3   Proposed Methodology for Indication of Bone Mineral Density

Researchers have investigated the relationship between bone health and bone X-ray images through various experiments. This study is based on the concept of transitive inference. Suppose R is a relation on set A. Relation R is transitive if $\forall x, y, z \in A, ((xRy) \bigwedge (yRz)) \Rightarrow (xRz)$. A transitive relation between premises is said to occur when the conclusion a R c (read as 'a related to c') follows logically from the premise relations a R b and b R c(a related to b and b related to c). Humans are capable of making transitive inferences and do so routinely. Transitive inference is a form of deductive reasoning. This work is based on the transitive inference: If bone mineral density is related to the DEXA scan of the person, and the DEXA scan can be related to the single X-ray scan, then bone mineral density can be related to the single X-ray scan.

We propose the following methodology: (1) In preprocessing, image filters and H-domes image slicing is applied to enhance contrast in X-ray images, (2) Textural features, statistical features and fractal dimension are computed in feature extraction and (3) three supervised algorithms namely, support vector machines, generalized boosted regression model and extreme gradient boosting are applied to classify patients to have bone mineral density to be high (Normal), or Abnormal (medium and low). The three classification algorithms are also applied to discriminate between Normal, Medium (osteopenia condition) and Low (osteoporosis condition). The three algorithms are applied on female patients alone to infer if distinction is more apparent in females.

Raw X-ray images possess low frequency noise due to X-ray diffusion in soft tissues and high frequency noise due to X-ray acquisition and Imaging characteristics. Chappard et al. [3,4] have applied a median filter to eliminate low frequency noise and a "top hat" filter to eliminate high frequency noise to increase the contrast of the bony structures within the image. Guggenbuhl et al. [9] binarized the images by applying automatic thresholding based on the histogram frequency distribution of grey levels. Abidi et al. [1] applied a series of common enhancement algorithms like linear regression, gamma intensity adjustment, logarithmic intensity adjustment, histogram equalization, edge and morphological operations etc. on X-ray images of luggage scenes and compared the results; They also proposed a new method to determine the optimal number of clusters or thresholds when segmenting X-ray images consisting of low density threat weapons. They used image hashing via intensity slicing such that objects of different intensity values are more visible. Intensity slicing can be done by equal interval image slicing, cumulative image slicing and H-domes image slicing. In this paper, we propose to use median filter to eliminate low frequency noise and "top hat" filter to eliminate high frequency noise and H-domes image slicing for intensity slicing of image hashing for the bone X-ray images.

Materka and Strzlecki [13] have discussed that image texture is a rich source of visual information and that feature extraction, texture discrimination, texture classification and shape reconstruction are the major issues in texture analysis; Results from the feature extraction stage are used for texture discrimination, texture classification or object shape determination. Also, to be noted is that analysis of fractal dimension can additionally enrich diagnostic knowledge about bone microarchitecture. Chappard et al. [3,4], Guggenbuhl et al. [9] have used run-length distribution, "skyscraper" fractal analysis, "blanket" fractal analysis and statistical analysis in their experiments to establish the correlation between texture analysis and bone mineral content, between texture analysis and histomorphometry and between texture analysis and bone micro-CT, respectively. In this paper, we propose to use the texture features, run-length distribution, fractal dimension, and statistical parameters as features. Normally, bone density accumulates during childhood and reaches a peak by around 25 years of age. Bone density is then maintained for about 10 years. After the age of 35, both men and women will normally lose 0.3–0.5% of their bone density per year as part of the aging process. Also women are more susceptible to low bone density. Hence, age and gender are also added into the feature set.

Learning methods can be classified into supervised, unsupervised and reinforcement. Supervised learning methods (or Classifiers) exhibit good mapping capabilities between the input patterns and their associated outputs(classes). They learn by examples, can predict new outcomes from past trends and can process information in parallel at high speed and in a distributed manner. Data are partitioned into training and test sets. Classifiers are continuously made to learn the instances in the training set to map to their classes, upto an allowed tolerance and this is the training phase. In the testing phase, the test data are input to verify the class of the instances by the stabilized weights in the training

phase. Cross-validation is to split the training set into training set and validation set. A k-fold cross-validation partitions the training set into k sets. For each
model complexity, the learner learns k times, each time using one of the sets as
the validation set and the other sets as the training set. It selects the model that
has the smallest average error on the validation set.

The features so extracted are applied to three classification algorithms Support Vector Machines (SVM), Generalized Boosted Regression Model (GBM)
and Extreme Gradient Boosting (XGB). A 10-fold cross-validation was applied
to all the algorithms. Support vector machines (SVM) [18] are supervised learning models with associated learning algorithms that analyze data for classification and regression. Boosting refers to a general and provably effective method
of producing a very accurate prediction rule by combining rough and moderately inaccurate rules. It has good generalization properties and seems not to
overfit in practice. The most straight forward generalization to a multiclass case,
called AdaBoost is adequate when the weak learner is strong enough to achieve
reasonable high accuracy [20]. Boosting has a sound theoretical base and also
has the advantage that the extra computation time that it requires is known in
advance. Moreover, in some applications where the dataset is large, the improvement is dramatic, due to the reduction of classification errors and well worth the
computational cost [16]. The family of boosting methods is based on a different, constructive strategy of ensemble information. To establish a connection
with the statistical framework, a gradient-descent based formulation of boosting
methods was derived by Freund and Schapire [6] and Friedman et al. [7,8]. This
formulation of boosting methods and the corresponding models were called the
gradient boosting machines. This framework also provided the essential justifica-
tions of the model hyper parameters. The ensemble models are a useful practical
tool for different predictive tasks, as they can consistently provide higher accuracy results compared to conventional single strong machine learning models
[14]. Xgboost is an efficient and scalable implementation of gradient boosting
framework [5].

This paper proposes to use the above set of algorithms in R [17] for the
classification of bone mineral density and glean sufficient knowledge in order to
deduce the transitive inference.

4   Classification of Bone Mineral Density in X-Ray Images

The raw X-ray images are impregnant with noise due to diffusion in soft tissues
and acquisition and imaging characteristics. So a variety of preprocessing techniques need to be applied to each X-ray image before features are extracted.
Extraction of essential features is a necessary step in a supervised learning
process so as to differentiate the classes appropriately. The X-ray bone image
undergoes preprocessing, feature extraction and classification phases.

4.1 Preprocessing of X-Ray Images

The raw X-ray image is subjected to Weiner filter to reduce noise. The background effects are removed by top hat and bottom hat filters which also have the benefit of adding a significant degree of smoothing to the spectrum. H-domes image slicing is further applied to remove small grains and this slicing method was chosen over equal interval image slicing and cumulative image slicing based on the peak signal to noise ratio (PSNR) on a subset of 25 bone images. The PSNR value is a quantitative measure that is used to compare filter performance between original and the reconstructed image. Figure 2a represents a sample original pelvis image of a patient and 2b represents the enhanced image.

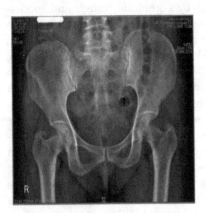 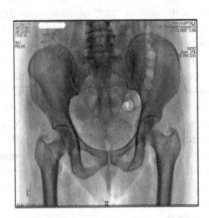

Fig. 2. Pelvis X-ray image: a. original b. H-domes enhanced

4.2 Feature Extraction

Feature extraction is a process by which the X-ray image is broken down into components of essential information that are more useful for describing its structure. The histogram h(i) contains the first order statistical information about the image, i = 0, 1,.. G − 1 where G is the total number of intensity levels in the image. h(i) contains the number of pixels in the whole image, which has this intensity. The approximate probability density of occurrence of the intensity levels is given by

$$p(i) = h(i)/M * N \tag{1}$$

where the size of the image is M × N. First-order statistical properties or central moments are derived from it to characterize the texture. The mean takes the average level of intensity and is defined as

$$mean = \mu = \Sigma_{i=o}^{G-1}(i * p(i)) \tag{2}$$

The variance describes the variation of intensity around the mean and is defined as

$$variance = \sigma^2 = \Sigma_{i=o}^{G-1}(i - \mu)^2 * p(i)) \tag{3}$$

Energy is defined as

$$Energy(E) = \Sigma_{i=o}^{G-1}(p(i))^2 \tag{4}$$

Entropy is a measure of histogram uniformity and is defined as

$$Entropy(H) = -\Sigma_{i=o}^{G-1}(p(i) * log_2(p(i))) \tag{5}$$

These first order measures cannot completely describe texture. The second order histogram is defined as the gray level co-occurrence matrix (GLCM) (joint probability distributions of pair of pixels) [10]. Contrast (or inertia) returns a measure of the intensity contrast between a pixel and its neighbor over the whole image and is defined as

$$Contrast = \Sigma_{i=o}^{G-1}\Sigma_{i=o}^{G-1}(i-j)^2 * p(i,j) \tag{6}$$

Correlation returns a measure of how correlated a pixel is to its neighbor over the whole image and is defined as

$$Correlation = \frac{\Sigma_{i=o}^{G-1}\Sigma_{i=o}^{G-1}(i-\mu_i)(j-\mu_j)p(i,j)}{\sigma_i\sigma_j} \tag{7}$$

where μ_i, μ_j, σ_i and σ_j denote the mean and standard deviations of the row and column sums of the matrix. Energy is calculated from GLCM as

$$Energy = \Sigma_{i=o}^{G-1}\Sigma_{i=o}^{G-1}p(i,j)^2 \tag{8}$$

Homogeneity returns a value that measures the closeness of the distribution of elements in the GLCM to the GLCM diagonal and is defined as

$$Homogeneity = \Sigma_{i=o}^{G-1}\Sigma_{i=o}^{G-1}\frac{p(i,j)}{1+(i-j)^2} \tag{9}$$

A feature vector is created for each X-ray image based on global qualities like average gray level, average contrast, smoothness, moments, uniformity and

Table 1. Feature extraction

Algorithm I Feature Extraction
1: for each image in the X-ray set do
2:       Apply Weiner filter
3:       Apply top and bottom hat filtering
4:       Apply hDome Image enhancement
5:       Get texture features of the image
6:       Get properties from gray level co-occurrence matrix
7:       Get runlength features and fractal dimension
8:       Include age and gender as features
10: end for
The feature values are stored

entropy (Eqs. 1 to 5), gray level run length statistics like Short Run Emphasis (SRE), Gray Level Run Emphasis (GRE), Gray Level Non-uniformity (GLN), Run Percentage (RP), Run length non-uniformity (RLN), and fractal dimension, gray level co-occurrence matrix (GLCM) characteristics like contrast, correlation, energy and homogeneity (Eqs. 6 to 9). Age and gender are also added into the feature vector. Algorithm I in Table 1 states the procedure for feature extraction.

4.3   Classification

R [17] is a language and environment for statistical computing and graphics. It is a GNU project developed at Bell laboratories by John Chambers and colleagues. R provides a wide variety of statistical (linear and nonlinear modelling, classical statistical tests, time-series analysis, classification, clustering) and graphical techniques, and is highly extensible. R is designed around a true computer language and it allows users to add additional functionality by defining new functions. As specified above the supervised learning algorithms used in this paper are SVM, GBM and XGB. The results are tabulated in Table 2 for a 2 class problem. The ROC curves are specified in Figs. 3, 4 and 5 for SVM, GBM abd XGB respectively.

Table 2. Results for normal/abnormal classification (all patients)

Classifier	Accuracy %	Sensitivity	Specivicity	AUC
SVM	82.00	0.92	0.68	0.85
GBM	82.00	0.90	0.68	0.88
XGB	89.00	0.92	0.65	0.88

5   Results and Discussion

PSG Hospitals is a 810 bedded, tertiary care hospital with qualified and experienced faculty and state of the art infrastructure, affiliated to PSG Institute of Medical Sciences & Research, Coimbatore, India [15]. The hospital houses various specialties like Medicine, Surgery, Obstetrics & Gynecology, Paediatrics, Orthopaedics, Radiology and Emergency Medicine. The hospital has been recognized for implementation of various social welfare schemes and is an internationally acclaimed centre to conduct various clinical trials.

Experiments are conducted on 311 pelvis and 160 spine images of the patients screened over a three month period at PSG Hospitals. The X-ray images are 760×580 pixels in size. The age of the patients varied between 10 and 90 and the average age was 46.37. There are 253 males and 218 females. Three experienced doctors identified the class attribute seperately based on the X-ray image and the

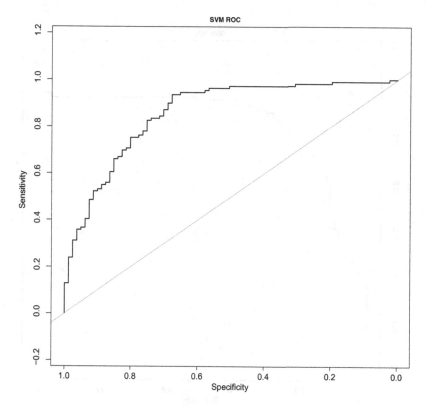

Fig. 3. ROC curve for SVM

class type (Normal/Osteopenia/Osteoporosis) was decided based on the majority voting by them. There was 25% of tie in the class attribute in such a process and lower the class attribute was decided with high density being normal, Osteopenia being medium and Osteoporosis being low.

After the images were pre-processed for enhancement, global, local statistical features, GLCM features, fractal dimension were extracted, age and gender were added and a total of 19 features are finally decided. The feature values are normalized. The 471 images are divided into training and testing at a 60:40 ratio at random and a 10-fold cross validation is used. For comparison purposes, the same set of data are classified by SVM, GBM and XGB. The results are tabulated in Table 2 for the two class problem (Normal/Abnormal) and in Table 3 for a three class problem (Normal/Osteopenia/Osteoporosis). It can be seen that the methodology identifies between normal and abnormal with an accuracy of 89% but when it is to distinguish between osteopenia and osteoporosis away from normal, the accuracy has reduced to 62%. Since, women were more susceptible to low bone density, the methodology was applied only to X-rays of female patients and found to have 82% accuracy. The results are tabulated in Table 4.

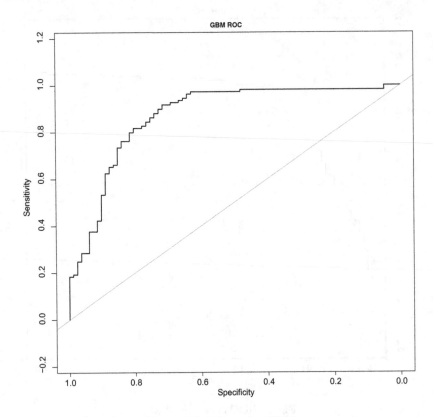

Fig. 4. ROC curve for GBM

Table 3. Results for normal/osteopenia/osteoporosis (all patients)

Classifier	Accuracy %	Sensitivity	Specivicity	AUC
SVM	57.00	0.55	0.78	0.55
GBM	62.00	0.60	0.82	0.55
XGB	59.00	0.55	0.76	0.57

Table 4. Results for normal/abnormal classification (only females)

Classifier	Accuracy %	Sensitivity	Specivicity	AUC
SVM	74.00	0.80	0.61	0.83
GBM	62.00	0.87	0.10	0.51
XGB	82.00	0.95	0.57	0.86

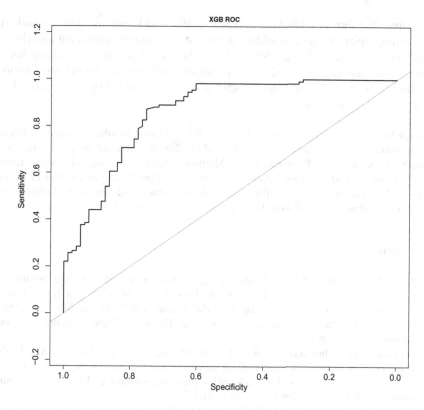

Fig. 5. ROC curve for XGB

6   Conclusion

The methodology described in this paper has extended the idea and relationship established by various researchers between bone mineral health and X-ray radiography images. DEXA equipment from different manufacturers might not give identical results, because of differences in calibration and bone edge detection algorithms. Moreover, the BMD values specified by WHO is based on white women [2] and as a result, the BMD values calculated for other countries or on other race for the same age cannot have this as the reference.

The proposed system is based on the strong correlation between the X-ray images and their underlying microarchitecture and the health of the bones. The system weighs on the transitive dependency and explores through inferential component, the bone mineral density by classification methods. The proposed system could identify abnormality with an accuracy of 89% for all patients and with an accuracy of 82% for only females, though the accuracy reduced to 62% for the three class problem. The proposed system is cheaper, portable, versatile and dependable to study changes in the density of minerals in X-ray images of bones. It is meant to be a guidance to the doctors to prescribe further medication if required. Similar to DEXA, it is a non-contact and non-invasive method. As more radiography images are added to the data set, the classification algorithms would be able to learn the characteristics of the images pertaining to a particular demography, age and race and able to classify much better. Additional features or a better image enhancement algorithm on the X-ray images may indeed be the future scope of work.

Acknowledgement. Authors wish to thank Dr. V Shyam Sundar, Former Professor, Department of Orthopaedics, PSG Institute of Medical Science and Research, currently Consultant Orthopaedic Surgeon, Kovai Medical Center and Hospital, Erode, India; Dr. M Arvind Kumar, Professor and S Udaymoorthy, Post Graduate Student, Department of Orthopaedics, PSG Institute of Medical Science and Research, Coimbatore, India for verifying and validating the class type in the data set.

References

1. Abidi, B., Mitckes, M., Abidi, M., Liang, J.: Grayscale enhancement techniques of x-ray images of carry-on luggage. In: Proceedings of SPIE 6th International Conference on Quality Control by Artificial Vision, vol. 5132, pp. 579–591 (2003)
2. Bone Densitometry. http://courses.washington.edu/bonephys/opbmd.html. Accessed 26 Aug 2016
3. Chappard, D., Chennebault, A., Moreau, M., Legrand, E., Audran, M., Basle, M.F.: Texture analysis of x-ray radiographs is a more reliable descriptor of bone loss than mineral content in a Rat model of localized disuse induced by the Clostridium botulinum toxin. Bone **28**(1), 72–79 (2001)
4. Chappard, D., Guggenbuhl, P., Legrand, E., Basle, M.F., Audran, M.: Texture analysis of x-ray radiographs is correlated with bone histomorphometry. J. Bone Miner. Metab. **23**, 24–29 (2005)
5. Chen, T., He, T.: xgboost: eXtreme Gradient Boosting. R package version 0.4-2 (2015)
6. Freund, Y., Schapire, R.: A decision-theoretic generalization of on-line learning and an application to boosting. J. Comput. Syst. Sci. **55**, 119–139 (1997)
7. Friedman, J.: Greedy boosting approximation: a gradient boosting machine. Ann. Stat. **29**, 1189–1232 (2001). doi:10.1214/aos/1013203451
8. Friedman, J., Hastie, T., Tibshirani, R.: Additive logistic regression: a statistical view of boosting. Ann. Stat. **28**, 337–407 (2000). doi:10.1214/aos/1016218222
9. Guggenbuhl, P., Bodic, F., Hamel, L., Basle, M.F., Chappard, D.: Texture analysis of x-ray radiographs of iliac bone is correlated with bone micro-CT. Osteoporos. Int. **17**, 447–454 (2006)
10. Haralick, R.M., Shanmugam, K., Dinstein, I.H.: Textural features for image classification. IEEE Trans. Syst. Man Cybern. **6**, 610–621 (1973)
11. Jimenez-Mendoza, D., Espinosa-Arbelaez, D.G., Giraldo-Betancur, A.L., Hernandez-Urbiola, M.I., Vargas-vazquez, D., Mario, E.: Single x-ray transmission system for bone mineral density determination. Rev. Sci. Instrum. **82**, 125105 (2011)
12. Luo, G., Kinney, J.H., Kaufman, J.J., Haupt, D., Chiabrera, A., Siffert, R.S.: Relationship between plain radiographic patterns and three-dimensional trabecular architecture in the human calcaneus. Osteoporos. Int. **9**, 339–345 (1999)

13. Materka A., Strzlecki M.: Texture analysis methods a review. COST B11 report, Institute of Electronics, Technical University of Lodz, Brussels (1998)
14. Natekin, A., Knoll, A.: Gradient boosting machines, a tutorial. Front. Neurorobotics **7**, 21 (2013). http://doi.org/10.3389/fnbot.2013.00021
15. PSG Institute of Medical Sciences & Research, Coimbatore, India. http://psgimsr. ac.in/hospitals.html. Accessed 12 Aug 2016
16. Ross, Q.J.: Bagging, boosting, and C4. 5. In: AAAI/IAAI, vol. 1 (1996)
17. R Project and Foundation. https://www.r-project.org/about.html. Accessed 12 Aug 2016
18. Vapnik, V.N.: The Nature of Statistical Learning Theory. Springer, Berlin (1995)
19. Veenland J.F.: Texture analysis of the radiographic trabecular bone pattern in osteoporosis. Ph.D. thesis, Reasmus University, Rotterdam (1999). ISBN 90-75655-04-5
20. Freund, Y., Scaphire, R.E.: Experiments with a new boosting algorithm. In: Proceedings of 13th International Conference in Machine Learning, pp. 148–156 (1996)

Author Index

Printed in the United States
by Baker & Taylor Publisher Services